IFCoLog Journal of Logic and its Applications

Volume 2, Number 1

May 2015

Disclaimer
Statements of fact and opinion in the articles in IfCoLog Journal of Logics and their Applications are those of the respective authors and contributors and not of the IfCoLog Journal of Logics and their Applications or of College Publications. Neither College Publications nor the IfCoLog Journal of Logics and their Applications make any representation, express or implied, in respect of the accuracy of the material in this journal and cannot accept any legal responsibility or liability for any errors or omissions that may be made. The reader should make his/her own evaluation as to the appropriateness or otherwise of any experimental technique described.

© Individual authors and College Publications 2015
All rights reserved.

ISBN 978-1-84890-178-0
ISSN (E) 2055-3714
ISSN (P) 2055 3706

College Publications
Scientific Director: Dov Gabbay
Managing Director: Jane Spurr

http://www.collegepublications.co.uk

Printed by Lightning Source, Milton Keynes, UK

All rights reserved. No part of this publication may be reproduced, stored in a retrieval system or transmitted in any form, or by any means, electronic, mechanical, photocopying, recording or otherwise without prior permission, in writing, from the publisher.

Editorial Board

Editors-in-Chief
Dov M. Gabbay and Jörg Siekmann

Marcello D'Agostino
Arnon Avron
Jan Broersen
Martin Caminada
Balder ten Cate
Agata Ciabttoni
Robin Cooper
Luis Farinas del Cerro
Didier Dubois
PM Dung
David Fernandez Duque

Jan van Eijck
Melvin Fitting
Michael Gabbay
Murdoch Gabbay
Wesley H. Holliday
Shalom Lappin
George Metcalfe
David Pearce
Henri Prade
David Pym
Ruy de Queiroz
Ram Ramanujam

Chrtian Retoré
Ulrike Sattler
Jörg Siekmann
Jane Spurr
Kaile Su
Leon van der Torre
Yde Venema
Heinrich Wansing
Jef Wijsen
John Woods
Michael Wooldridge

Scope and Submissions

This journal considers submission in all areas of pure and applied logic, including:

- pure logical systems
- proof theory
- constructive logic
- categorical logic
- modal and temporal logic
- model theory
- recursion theory
- type theory
- nominal theory
- nonclassical logics
- nonmonotonic logic
- numerical and uncertainty reasoning
- logic and AI
- foundations of logic programming
- belief revision
- systems of knowledge and belief
- logics and semantics of programming
- specification and verification
- agent theory
- databases
- dynamic logic
- quantum logic
- algebraic logic
- logic and cognition
- probabilistic logic
- logic and networks
- neuro-logical systems
- complexity
- argumentation theory
- logic and computation
- logic and language
- logic engineering
- knowledge-based systems
- automated reasoning
- knowledge representation
- logic in hardware and VLSI
- natural language
- concurrent computation
- planning

This journal will also consider papers on the application of logic in other subject areas: philosophy, cognitive science, physics etc. provided they have some formal content.

Submissions should be sent to Jane Spurr (jane.spurr@kcl.ac.uk) as a pdf file, preferably compiled in LaTeX using the IFCoLog class file.

CONTENTS

Editorial comment about "On the Difference between ABA and AA" 1
  Dov Gabbay

ARTICLES

On the Difference between Assumption-Based Argumentation and Abstract
  Argumentation 15
  Martin Caminada, Samy Sá, João Alcântara and Wolfgang Dvořák

Physarum Polycephalum Diagrams for Syllogistic Systems 35
  Andrew Schumann and Andrew Adamatzky

Tabdual: a Tabled Abduction System for Logic Programs 69
  Ari Saptawijaya and Luís Moniz Pereira

FORTHCOMING PAPERS

Cut-free Proof Systems for Geach Logics 125
  Melvin Fitting

Meta-level Abduction 127
   Katsumi Inoue

Abduction and Beyond 129
   Ari Saptawijaya and Luís Moniz Pereira

EDITORIAL COMMENT ABOUT "ON THE DIFFERENCE BETWEEN *ABA* AND *AA*"

DOV GABBAY
FLAP Editor-in-Chief
dov.gabbay@kcl.ac.uk

I am writing this editorial about the paper of M. Caminada, S. Sá, J. Alcântara and W. Dvořák, appearing in the current issue. I felt the need for a clarifying editorial discussing some possible misunderstandings in the argumentation community about what is assumption based argumentation (ABA) and its relation to abstract argumentation (AA).

The debate/controversy generated by the current paper of Caminada *et al.* [11] directly relates to the current misunderstanding, and prompted this editorial.

It seems that the ABA community would like to think that ABA is equivalent to AA, while the current paper seems to give a counter-example to that. There has been extensive back and forth communications about the validity of this counter-example between the authors, the referees and senior members of the ABA community. I hope that this editorial will help clarify the source of confusion. The perceptive reader will note that this editorial contains technical results. Unfortunately, this was necessary, because sometimes to make an editorial point, one needs to prove it!

1 What are the target objects of attacks involved in the ABA approach? Are they single assumptions or are they sets of assumptions?

Let our starting point of departure be the notion of an ABA framework. We can use Definition 2 of the Caminada *et al.* paper in this issue. There is no misunderstanding in the community about this notion.

In order to make this editorial independent, let me reproduce here the text from the Caminada *et al.* paper [11].

> Begin quote:
>
> Over the years, different versions of the assumption-based argumentation framework have become available [9, 10, 8] and these versions give slightly

different formalizations. For current purposes, we apply the formalization described in [8] which not only is the most recent, but is also relatively easy to explain.

Definition 1 ([8]). *Given a deductive system $\langle \mathcal{L}, \mathcal{R} \rangle$ where $\mathcal{L}$ is a logical language and $\mathcal{R}$ is a set of inference rules on this language, and a set of assumptions $\mathcal{A} \subseteq \mathcal{L}$, an* argument *for $c \in \mathcal{L}$ (the* conclusion *or* claim*) supported by $S \subseteq \mathcal{A}$ is a tree with nodes labelled by formulas in $\mathcal{L}$ or by the special symbol $\top$ such that:*
- *the root is labelled c*
- *for every node N*
    - *if N is a leaf then N is labelled either by an assumption (in S) or by $\top$[1]*
    - *if N is not a leaf and b is the label of N, then there exists an inference rule $b \leftarrow b_1, \ldots, b_m$ ($m \geq 0$) and either $m = 0$ and the child of N is labelled by $\top$, or $m > 0$ and N has m children, labelled by $b_1, \ldots, b_m$ respectively*
- *S is the set of all assumptions labelling the leaves*

We say that a set of assumptions $\mathcal{A}sms \subseteq \mathcal{A}$ enables the construction of *an argument A (or alternatively, that A* can be constructed based on $\mathcal{A}sms$*) if A is supported by a subset of $\mathcal{A}sms$.*

Notice that each assumption $\alpha \in \mathcal{A}$ enables an argument A_α with claim α supported by $\{\alpha\}$. That is, the corresponding tree has just one node that is labeled α.

Definition 2 ([8]). *An ABA framework is a tuple $\langle \mathcal{L}, \mathcal{R}, \mathcal{A}, \bar{\ } \rangle$ where:*
- *$\langle \mathcal{L}, \mathcal{R} \rangle$ is a deductive system*
- *$\mathcal{A} \subseteq \mathcal{L}$ is a (non-empty) set, whose elements are referred to as assumptions*
- *$\bar{\ }$ is a total mapping from $\mathcal{A}$ into $\mathcal{L}$, where $\bar{\alpha}$ is called the contrary of α*

For current purposes, we restrict ourselves to ABA-frameworks that are *flat* [9], meaning that no assumption is the head of an inference rule. Furthermore, we follow [8] in that each assumption has a unique contrary.

---
[1]Professor F. Toni, who commented on a draft of this editorial, asked me to stress that "$\top$ is used to separate facts from assumptions in the definition of arguments as trees. It is not an assumption, and, indeed, we impose that it does not belong to $\mathcal{L}$ (and $\mathcal{A} \subseteq \mathcal{L}$)"

We are now ready to define the various abstract argumentation semantics (in the context of an ABA-framework). We say that an argument A_1 *attacks* an argument A_2 iff the conclusion of A_1 is the contrary of an assumption in A_2. Also, if $Args$ is a set of arguments, then we write $Args^+$ for $\{A \mid$ there exists an argument in $Args$ that attacks $A\}$. We say that a set of arguments $Args$ is *conflict-free* iff $Args \cap Args^+ = \emptyset$. We say that a set of arguments $Args$ *defends* an argument A iff each argument that attacks A is attacked by an argument in $Args$.

⋮

The next step is to describe the various ABA semantics. These are defined not in terms of sets of arguments (as is the case for abstract argumentation) but in terms of sets of assumptions. A set of assumptions $Asms_1$ is said to *attack* an assumption α iff $Asms_1$ enables the construction of an argument for conclusion $\overline{\alpha}$. A set of assumptions $Asms_1$ is said to attack a set of assumptions $Asms_2$ iff $Asms_1$ attacks some assumption $\alpha \in Asms_2$. Also, if $Asms$ is a set of assumptions, then we write $Asms^+$ for $\{\alpha \in \mathcal{A} \mid Asms$ attacks $\alpha\}$. We say that a set of assumptions $Asms$ is *conflict-free* iff $Asms \cap Asms^+ = \emptyset$. We say that a set of assumptions *defends* an assumption α iff each set of assumptions that attacks α is attacked by $Asms$.
End quote.

Let me ask the reader a simple question. The above quote from the Caminada *et al.* paper defined what it means for one set of assumptions to attack another.

My question is:

- What is the set S of objects which are targeted/attacked in the ABA approach? Is it the set of assumptions $\mathcal{A}$ or is it all the non-empty subsets of $\mathcal{A}$? In other words, is $S = \mathcal{A}$ or is $S = (2^{\mathcal{A}} - \emptyset)$?[2]

This is exactly the source of misunderstanding. The reader might think that it must be $S = (2^{\mathcal{A}} - \emptyset)$, because we are talking about sets of assumptions attacking other sets of assumptions, but this is not enough to decide that $S = (2^{\mathcal{A}} - \emptyset)$. A formal standard

---

[2]Again, I am grateful to Professor F. Toni for the following comment on an earlier draft of this Editorial "We never said in any paper that arguments are sets of assumptions, they are not. ABA can be defined in terms of sets of assumptions or in terms of arguments which are deductions supported by sets of assumptions (in different papers we use different notions of deductions, but they all boil down to the same, semantically, as we prove). The two views (sets of assumptions vs sets of arguments) coincide (in a way that we specify) for all semantics we defined for ABA.

We never restrict arguments to have a non-empty set of assumptions as support, in general or in any of the instances we studied."

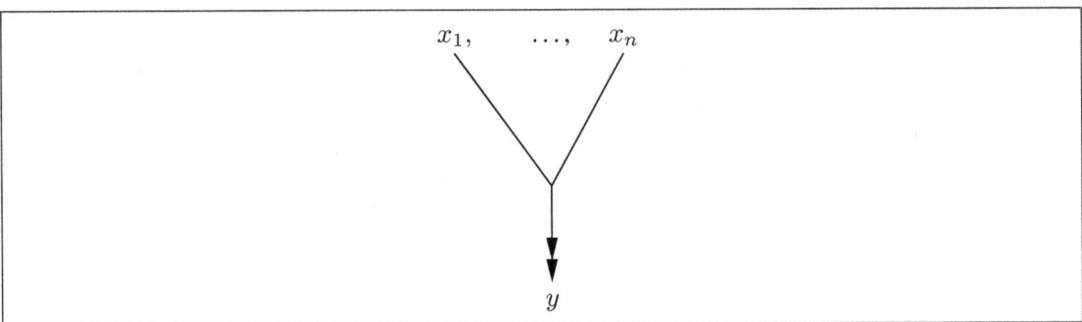

Figure 1

definition must be given and such an agreed definition does not exist in the literature. In fact, some papers use $S = \mathcal{A}$ and some papers use $S = (2^{\mathcal{A}} - \varnothing)$.

We now explain the technical side of this point.

An examination of the definitions above of sets of assumptions attacking other sets of assumptions show that the following observations (1), (2) and (3) hold:

1. A set of assumptions $E_1 \subseteq \mathcal{A}$ attacks E_2 exactly when for some $x \in E_2$ we have that E_1 attacks x.

2. It is possible that E_2 attacks any assumptions y but for no proper subset $E' \subsetneq E_2$ do we have that E' attacks y.

3. In any complete extension a set $\{e_1, \ldots, e_n\}$ is in the extension iff all of $\{e_i\}, i = 1, \ldots, n$ are in the extension.

Note that condition (3) is a consequence of condition (1).

Assume that $\{e_1, \ldots, e_n\}$ is in the extension but $\{e_i\}$ is not in the extension. Then $\{e_i\}$ is attacked by some X which is either in the extension or und. But then X attacks $\{e_1, \ldots, e_n\}$ and so X must be out, a contradiction. Assume that all $\{e_i\}$ are in the extension. Let X attack $\{e_1, \ldots, e_n\}$. By condition (1) X must attack some $\{e_i\}$. But $\{e_i\}$ is in, so X is out. Thus all attackers of $\{e_1, \ldots, e_n\}$ are out so $\{e_1, \ldots, e_n\}$ is in the extension.

Technically, whenever we have sets attacking sets which satisfy (1), (2) and (3), we can represent the situation faithfully using $S = \mathcal{A}$ and using the notion of joint attacks. See Figure 1. See [4, 5, 6].

We now discuss this point because, I believe, it will help explain the misunderstandings concerning ABA.

Definition 3.

1. Let S be a non-empty set. Let $R \subseteq (2^S - \emptyset) \times S$. Then (S, R) is a joint attack network. the relation GRx between non-empty subsets G of S and nodes $x \in S$ is the joint attack relation.

2. A labelling function $\lambda : S \mapsto \{in, out, und\}$ is a legitimate Caminada–Gabbay labelling (for networks with joint attacks) iff the following holds:

(CG1) $\lambda(x) = out$ iff for some G such that GRx we have that of all $y \in G, \lambda(y) = in$.

(CG2) $\lambda(x) = in$, iff for all G such that GRx, we have that there exists a $y \in G$ such that $\lambda(y) = out$.

(CG3) $\lambda(x) = und$ iff for all G such that GRx we have that there is a $y \in G$ with $\lambda(y) \neq in$ and furthermore, there is a G' such that $G'Rx$ and for all $y \in G', \lambda(y) \neq out$ and there is a $z \in G'$ such that $\lambda(z) = und$.

3. The complete extensions E of (S, R) are defined by the legitimate labellings λ

$$E_\lambda = \{x \in S | \lambda(x) = in\}.$$

Theorem 4. *Let S be a set and let $\mathcal{N}$ be an argumentation network based on the set $A = (2^S - \emptyset)$ as arguments and the attack relation $\mathbb{R} \subseteq A \times A$. Assume the following holds:*

1. *$X\mathbb{R}Y$ iff $\exists y \in Y(X\mathbb{R}\{y\})$*

2. *$X\mathbb{R}Y$ iff for some $X' \subseteq X(X'\mathbb{R}Y)$.*

3. *For any complete extension $\mathbb{E}$ of $(A, \mathbb{R})$ and any $X \subseteq A$ we have*

$$X \in \mathbb{E} \text{ iff } \forall x \in X(\{x\} \in \mathbb{E}).$$

Define the following joint attacks R on S, for $X \in A, y \in S$:

- *XRy iff $X\mathbb{R}\{y\}$.*

  Then the following holds:

- *$\mathbb{E}$ is a complete extension of $(A, \mathbb{R})$ iff $E = \{x | \{x\} \in \mathbb{E}\}$ is a complete extension of (S, R).*

Proof.

1. Assume $\mathbb{E}$ is a complete extension of $(A, \mathbb{R})$. $(A, \mathbb{R})$ is an ordinary traditional AA network. Therefore there exists a legitimate Caminada labelling $\mu : A \mapsto \{\text{in, out, und}\}$ such that $\mathbb{E} = \{a \in A | \mu(\{a\}) = \text{in}\}$. We define a legitimate labelling λ on (S, R) using μ as follows:
$$\lambda(x) = \text{def. } \mu(\{x\}), \text{ for } x \in S.$$

We prove that λ is a legitimate labelling.

(CG1) Assume $\lambda(x) = \text{out}$. This means that $\mu(\{x\}) = \text{out}$. So for some G such that $G\mathbb{R}\{x\}$ we have $\mu(G) = \text{in}$. By property (3) we have $\mu(\{y\}) = \text{in}$ for all $y \in G$. So we found a G such that GRx and $\lambda(y) = \text{in}$ for all $y \in G$.

(CG2) Suppose $\lambda(x) = \text{in}$. Then we have $\mu(\{x\}) = \text{in}$. Let G be any set such that GRx. Thus we have $G\mathbb{R}\{x\}$. Here $\mu(G) = \text{out}$. Thus by property (3), for some $y \in G$ we have $\mu(\{y\}) = \text{out}$, i.e. $\lambda(y) = \text{out}$. So we showed that if $\lambda(x) = \text{in}$, then all G such that GRx, there is a $y \in G$ such that $\lambda(y) = \text{out}$.

(CG3) Suppose $\lambda(x) = \text{und}$. So $\mu(\{x\}) = \text{und}$. So for any G such that $G\mathbb{R}\{x\}$ holds we have that either $\mu(G) = \text{out}$ or $\mu(G) = \text{und}$, with at least one $G'\mathbb{R}\{x\}$ with $\mu(G') = \text{und}$.

If $u(G) = \text{out}$, then for some $y \in G, \lambda(y) = \text{out}$. This we have shown in the (CG2) case.

If $\mu(G') = \text{und}$ then we cannot have $\mu(\{y\}) = \text{in}$ for all $y \in G'$ and we cannot have $\mu(\{y\}) = \text{out}$ for any $y \in G'$. So we have that for all $y \in G', \lambda(y) \neq \text{out}$ and for some $z \in G', \lambda(z) = \text{und}$.

This shows that (CG3) holds.

2. Let λ be a legitimate Caminada–Gabbay labelling of the joint attacks network (A, R). Define μ for $(A, \mathbb{R})$ as follows:

(μ_1) $\mu(\{e_1, \ldots, e_n\}) = \text{in}$, iff for all $i, \lambda(e_i) = \text{in}$.

(μ_2) $\mu(\{e_1, \ldots, e_n\}) = \text{out}$ iff for some $i, \lambda(e_i) = \text{out}$.

(μ_3) $\mu(\{e_1, \ldots, e_n\}) = \text{und}$ iff for all $i, \lambda(e_i) \neq \text{out}$ and for some $j, \lambda(e_j) = \text{und}$.

We show that μ is a legitimate labelling for $(A, \mathbb{R})$.

1. Assume $\mu(\{e_1, \ldots, e_n\}) = \text{in}$ and let $G\mathbb{R}\{e_1, \ldots, e_n\}$. Then for some i, GRe_i and since $\lambda(e_i) = \text{in}$, we get $\lambda(y) = \text{out}$ for some $y \in G$ and so $\mu(G) = \text{out}$.

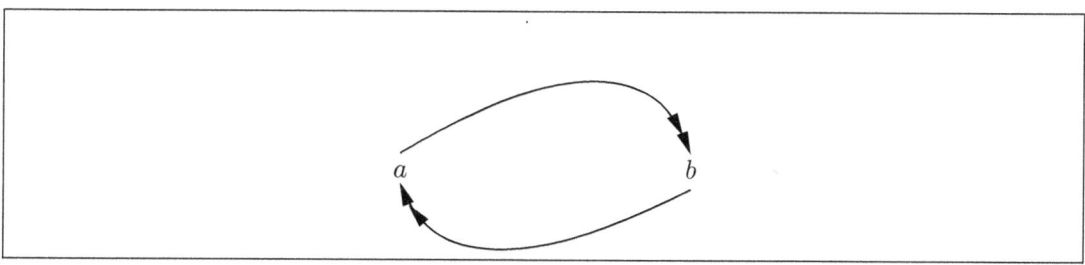

Figure 2

2. Assume $\mu(\{e_1,\ldots,e_n\}) = $ out. Then for some $i, \lambda(e_i) = $ out. Hence for some G, GRe_i and $\lambda(y) = $ in for all $y \in G$. Hence $\mu(G) = $ in, and of course we do have $G\mathbb{R}\{e_1,\ldots,e_n\}$.

3. Assume $\mu(\{e_1,\ldots,e_n\}) = $ und. Then for some $j, \lambda(e_j) = $ und and for all $i, \lambda(e_i) \neq $ out.

   Without loss of generality, we can assume therefore that say $\lambda(e_1),\ldots,\lambda(e_k) = $ und, $1 \leq k < n$, and $\lambda(e_{k+1}),\ldots,\lambda(e_n) = $ in. Let $G\mathbb{R}\{e_1,\ldots,e_n\}$. Then $G\mathbb{R}\{e_j\}$ for some j. So GRe_j holds.

   If $j \leq k$, then for some $y \in G, \lambda(y) = $ und and for all $x \in G, \lambda(x) \neq $ out and if $j > k$ then for some $y \in G, \lambda(y) = $ out.

   The above holds for any G attacking $\{e_1,\ldots,e_n\}$. But this is exactly the condition of $\mu(\{e_1,\ldots,e_n\}) = $ und in $(A,\mathbb{R})$.

   $\square$

Remark 1. *Definition 3 and Theorem 4 actually say that $(A,\mathbb{R})$ and (S,R) are the same. The difference is whether we count the subsets $G \subseteq S$ as separate objects to be attacked or not. To sharpen this point, look at the assumption based system $\{a, \bar{a}, b, \bar{b}\}$ with $a \vdash \bar{b}$ and $b \vdash \bar{a}$. This gives rise to the assumption based network of Figure 2.*

We can regard this as a network with joint attacks (except that there are no "real" joint attacks). So we take the arguments/objects to be attacked as $\{a,b\}, \{a\}, \{b\}$, with $\{a,b\}\mathbb{R}\{a\}, \{a,b\}\mathbb{R}\{b\}, \{a,b\}\mathbb{R}\{a,b\}, \{a\}\mathbb{R}\{b\}$ and $\{b\}\mathbb{R}\{a\}$.

This can be represented in Figure 3. Or, equivalently, according to Theorem 4 in Figure 4

Clearly what we have achieved here is *duplication of points/objects to be attacked!*
This clarification will help us understand the Caminada *et al.* example.

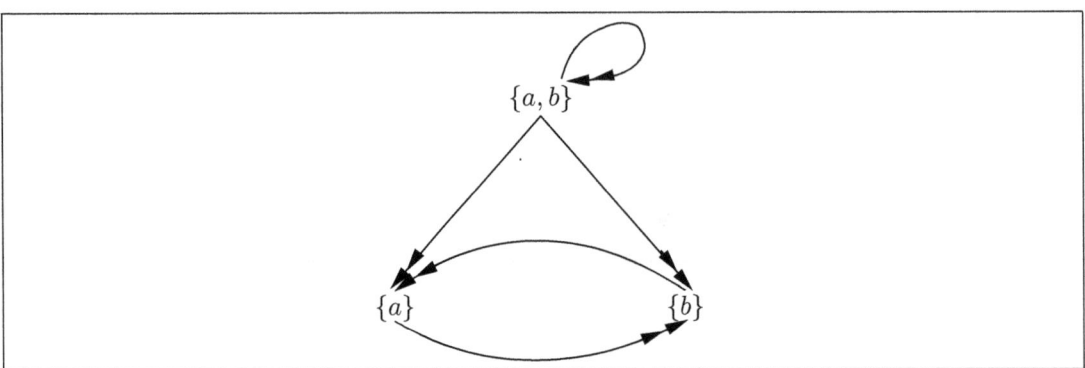

Figure 3

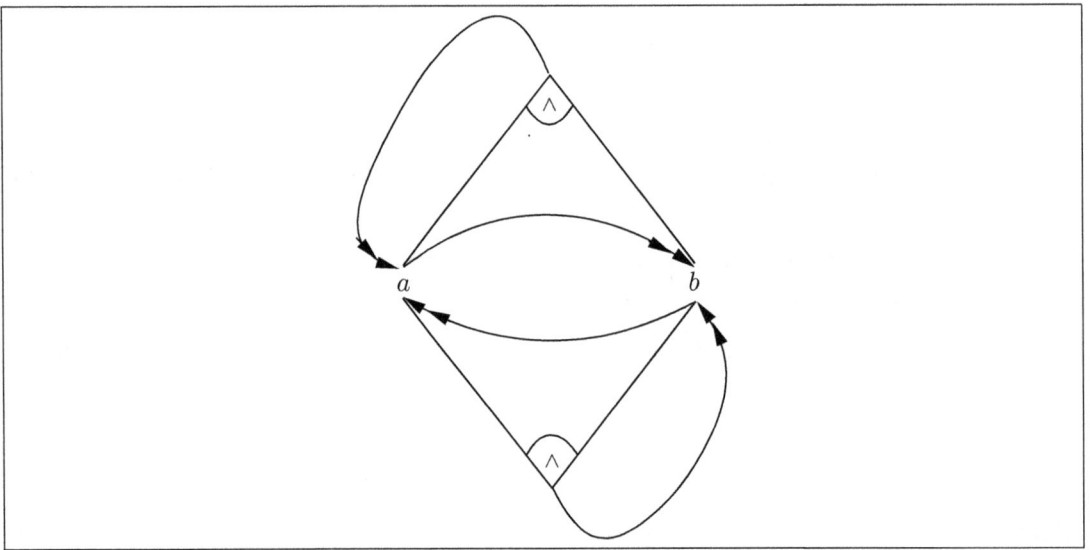

Figure 4

2 Examination of the Caminada *et al.* counter-example

We are now ready to examine the Caminada *et al.* example, as given in the proof of Theorem 7 of their paper.

The Caminada *et al.* counter-example has the assumption set $\{\alpha, \beta, \gamma, \varepsilon\}$ with the contraries (respectively) $\{a, b, c, d\}$. The attacks are derived from the following provability

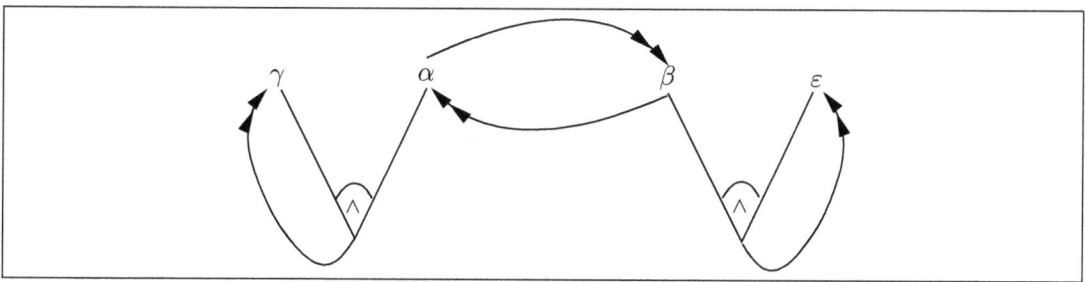

Figure 5

items:

$$\{\gamma\} \vdash c$$
$$\{\beta\} \vdash a$$
$$\{\alpha\} \vdash b$$
$$\{\gamma, \alpha\} \vdash c$$
$$\{\varepsilon, \beta\} \vdash e$$

The above sets are the key attacking elements involved.

In this example, the set $\mathcal{A}$ of assumptions is $\mathcal{A} = \{\alpha, \beta, \gamma, \varepsilon\}$.

If we use $\twoheadrightarrow$ to denote attack, and let $S = \mathcal{A} = \{\alpha, \beta, \gamma, \varepsilon\}$, we have:

$$\{\alpha\} \twoheadrightarrow \beta$$
$$\{\beta\} \twoheadrightarrow \alpha$$
$$\{\gamma\} \twoheadrightarrow \gamma$$
$$\{\beta, \varepsilon\} \twoheadrightarrow \varepsilon$$
$$\{\alpha, \gamma\} \twoheadrightarrow \gamma$$

The extensions are:

Ext 1: $\alpha = \beta = \gamma = \varepsilon = $ und.
Ext 2: $\beta = $ in, $\alpha = $ out, $\gamma = \varepsilon = $ und.
Ext 3: $\alpha = \varepsilon = $ in, $\beta = $ out, $\gamma = $ und.

It is clear that **Ext 2** is not semi-stable, because **Ext 3** has more "in–out" nodes.

We now turn to the other option, of $S = (2^{\mathcal{A}} - \varnothing)$. Before we get into detail, let us look at an intermediate option. Call it "option intermediate". In this intermediate case, we take for S only sets of assumptions supporting proof items, namely we let

$$S = \{\{\alpha\}, \{\beta\}, \{\gamma\}, \{\varepsilon\}, \{\alpha, \gamma\}, \{\varepsilon, \beta\}\}.$$

These are actually the assumption sets involved in the attacks of the Caminada *et al.* example. Figure 6 describes the argumentation network we get:

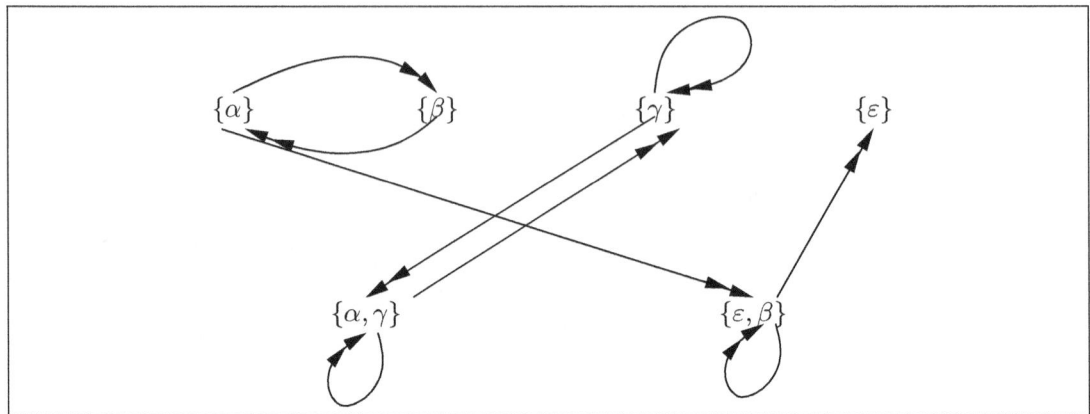

Figure 6

The extensions are:
Ext $i-1$: All undecided
Ext $i-2$: $\{\beta\}$ = in.
$\{\alpha\} = \{\alpha, \gamma\}$ = out
$\{\gamma\} = \{\varepsilon, \beta\} = \{\varepsilon\}$ = undecided.
Ext $i-3$: $\{\alpha\} = \{\varepsilon\}$ = in
$\{\beta\} = \{\varepsilon, \beta\}$ = out
$\{\alpha, \gamma\} = \{\gamma\}$ = undecided.

Notice that condition (3) holds!

In this case **Ext** $i-2$ is indeed semi-stable.

We now move to the case of all subsets making use of the intermediate ones.

If we let $S = (2^{\mathcal{A}} - \varnothing)$ and use the notation $\{x_i\}$ to denote the set containing x_i and $\{\{+x_i, -y_j\}\}$ to denote the family of all subsets $E \subseteq \mathcal{A}$ such that E contains x_i but does not contain any of y_j. Then Figure 7 represents the situation using our abbreviation.

The extensions are derived from the intermediate case Figure 6.

Ext 1*: all undecided
Ext 2*: $\{\{+\beta, -\alpha\}\}$ = in
$\{\{+\alpha\}\}$ = out
$\{\{+\gamma, -\alpha\}\} = \{\{+\varepsilon\}\}$ = undecided
Ext 3*: $\{\{+\alpha, -\beta, -\gamma\}\}$ = in
$\{\{+\varepsilon, -\beta, -\gamma\}\}$ = in
$\{\{+\beta\}\}$ = out.

Again, we get that **Ext 2*** is semi-stable and the counter-example does not work.

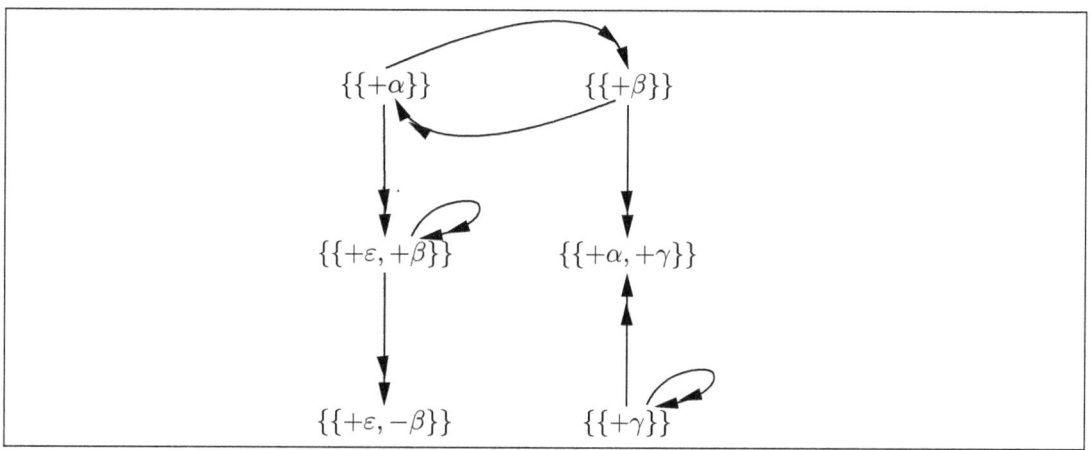

Figure 7

Conclusion

The counter-example depends for its success on the lack of duplication of arguments. It works for the case $S = \mathcal{A}$ and does not work for the case $S = (2^{\mathcal{A}} - \varnothing)$. Of course, there may still be a counter-example for the case $2^{\mathcal{A}} - \varnothing$.

Furthermore, we can wonder, if the reason the counter-example does not work for the case $S = 2^{\mathcal{A}} - \varnothing$ is just because of the possibility of duplication of points, then is this really important?

Caminada's point of view about his counter-example can be explained using Figure 5. He regards the arrows as the objects which attack each other in the spirit of the attack semantics of [12]. one arrow $G \twoheadrightarrow x$ attacks another arrow $H \twoheadrightarrow y$ if $x \in H$. Caminada compares this attack semantics with the semantics of the nodes/assumptions themselves.

We do not accept this view but the counter-example works for both views, his and ours.

The next section will evaluate the situation in detail.

3 Editorial Comment

My impression is that the discussion about the counter-example in the community went along social directions and missed the important technical points. The ABA community is keen on the idea that ABA = AA and did not like the counter-example and so argued against it saying (1) and (2) below:

1. Use $S = (2^{\mathcal{A}} - \varnothing)$ rather than use $S = \mathcal{A}$.

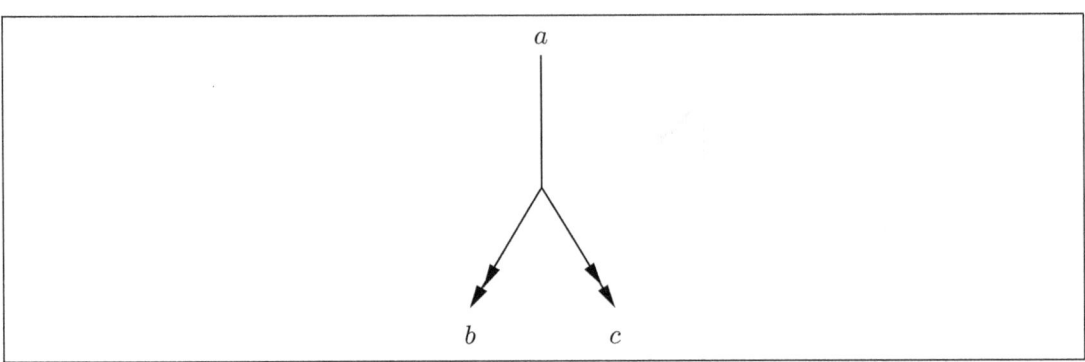

Figure 8

The facts are, however, that some key papers from the ABA community do use $S = \mathcal{A}$. See, for example, [1].

Also, the case of $S = \mathcal{A}$ is the obvious simple way to model ABA given that condition 1 holds, namely

- A set of assumptions X attacks a set Y iff it attacks a single $y \in Y$.

2. The semi-stable semantics is not important, it is marginal and so for all important semantics ABA=AA.

   Personally I do not buy this counterargument. Firstly because it is generic, if x gets in your way you can also say x does not count. Secondly if we accept that stable semantics is important then semi-stable is the next best one available, and should also be important

In this defensive mindset, they overlooked the importance of condition 1.

This is a crucial assumption because the real problem of importance is the crucial case of when it does not hold. What is important to investigate in the abstract is sets genuinely attacking sets. Consider the following example:

Assumptions $\{a, b, c\}$ contraries $\{\bar{a}, \bar{b}, \bar{c}\}$ respectively.

Rules
$$b, c \vdash a$$
$$a \vdash \bar{a}$$

here we have that a attacks $\{b, c\}$. ABA does not deal with this. ASPIC tries to deal with it. Technically we have here what I called in 2009 a disjunctive attack. See Figure 8

If a is in we want one of $\{b, c\}$ to be out.

I therefore propose the notion of set ABA (call it *SABA*), where we take $S = (2^A - \varnothing)$ and allow for sets of assumptions to attack sets of assumptions, when we have

$$X \vdash \bar{x}$$
$$Y \vdash x$$

We let X attack Y. $\bar{x}$ is the contrary of x. x need not be an assumption.

For disjunctive attacks, see [5, 6, 7] and [13].

Acknowledgements

I am grateful to Martin Caminada and Francesca Toni for comments on this Editorial.

References

[1] Claudia Schulz and Francesca Toni. Logic Programming in Assumption-Based Argumentation Revisited — Semantics and Graphical Representation. *AAAI 2015*.

[2] Francesca Toni. A tutorial on assumption-based argumentation. In *Argumentation and Computation*, 5:1, 89-117, DOI: 10.1080/19462166.2013.869878

[3] Claudia Schulz and Francesca Toni. Justifying Answer Sets using Argumentation.

[4] S. Holbech Nielsen and S. Parsons. A generalization of Dung's Abstract Framework for Argumentation Arguing with Sets of Attacking Arguments. *Third International Workshop on Argumentation in Multiagent Systems*, Hakodate, Japan, 2006

[5] Dov Gabbay. Fibring Argumentation Frames. *Studia Logica*, 93(2-3): 231–295, 2009,

[6] Dov Gabbay and Michael Gabbay. The attack as strong negation, Part 1. To appear in *Logic Journal of the IGPL*.

[7] Dov Gabbay. Theory of Semi-Instantiation in Abstract Argumentation. To appear *Logica Universalis*.

[8] P.M. Dung, R.A. Kowalski, and F. Toni. Assumption-based argumentation. In Guillermo Simari and Iyad Rahwan, editors, *Argumentation in Artificial Intelligence*, pages 199–218. Springer US, 2009.

[9] A. Bondarenko, P.M. Dung, R.A. Kowalski, and F. Toni. An abstract, argumentation-theoretic approach to default reasoning. *Artificial Intelligence*, 93:63–101, 1997.

[10] P. M. Dung, P. Mancarella, and F. Toni. Computing ideal sceptical argumentation. *Artificial Intelligence*, 171(10-15):642–674, 2007.

[11] Martin Caminada, Samy Sá, João Alcântara and Wolfgang Dvořák. On the difference between assumption-based argumentation and abstract argumentation. *IfColog Journal of Logics and their Applications*, 2015.

[12] S. Villata, G. Boella and L .van der Torre. Attack Semantics for Abstract Argumentation, *IJCAI 2011*.

[13] D. Gabbay, Theory of Disjunctive attacks in Argumentation Networks, to appear in *Logic Journal of IGPL*.

ON THE DIFFERENCE BETWEEN ASSUMPTION-BASED ARGUMENTATION AND ABSTRACT ARGUMENTATION

MARTIN CAMINADA
University of Aberdeen

SAMY SÁ
Universidade Federal do Ceará

JOÃO ALCÂNTARA
Universidade Federal do Ceará

WOLFGANG DVOŘÁK
Universität Wien

Abstract

In the current paper, we re-examine the connection between abstract argumentation and assumption-based argumentation. These two formalisms are often claimed to be equivalent in the sense that (a) evaluating an assumption based argumentation framework directly with the dedicated semantics, and (b) first constructing the corresponding abstract argumentation framework and then applying the corresponding abstract argumentation semantics, produce the same outcome. Although this holds for several semantics, in this work we show that there exist well-studied admissibility-based semantics (semi-stable and eager) under which equivalence does not hold.

Keywords: Assumption-Based Argumentation, Abstract Argumentation, Semi-Stable Semantics

Parts of the current paper were presented at BNAIC 2013, but have not been published in formal proceedings.

1 Introduction

The 1990s saw some of the foundational work in argumentation theory. This includes the work of Simari and Loui [20] that later evolved into Defeasible Logic Programming (DeLP) [15] as well as the ground-breaking work of Vreeswijk [23] whose way of constructing arguments has subsequently been applied in the various versions of the ASPIC formalism [7, 18, 17]. Two approaches, however, stand out for their ability to model a wide range of existing formalisms for non-monotonic inference. First of all, there is the abstract argumentation approach of Dung [11], which is shown to be able to model formalisms like default logic, logic programming under stable and well-founded model semantics [11], as well as Nute's defeasible logic [16] and logic programming under the 3-valued stable model semantics [24]. Secondly, there is the assumption-based argumentation approach of Bondarenko, Dung, Kowalski and Toni [4], which is shown to model formalisms like default logic, logic programming under stable model semantics, auto epistemic logic and circumscription [4].

One of the essential differences between these two approaches is that abstract argumentation is argument-based. One uses the information in the knowledge base to construct arguments and to examine how these arguments attack each other. The different semantics are then defined on the resulting argumentation framework (the directed graph in which the nodes represent arguments and the arrows represent the attack relation). In assumption-based argumentation, on the other hand, the definitions of the different semantics are based not on arguments but on sets of assumptions that attack each other based on their possible inferences.

One claim that occurs several times in the literature is that abstract argumentation and assumption-based argumentation are somehow equivalent. That is, the outcome (in terms of conclusions) of abstract argumentation would be the same as the outcome of assumption-based argumentation [12, 18]. In the current paper, we argue that although this equivalence does hold under *some* semantics, it definitely does not hold under *every* semantics. In particular, we show that under two well-known and well-studied admissibility-based semantics (semi-stable [22, 5, 8] and eager [6, 1, 13]) the outcome of assumption-based argumentation is fundamentally different from the outcome of abstract argumentation.

2 Preliminaries

Over the years, different versions of the assumption-based argumentation framework have become available [4, 10, 12] and these versions give slightly different formalizations. For current purposes, we apply the formalization described in [12] which not only is the most recent, but is also relatively easy to explain.

Definition 1 ([12]). *Given a deductive system $\langle \mathcal{L}, \mathcal{R} \rangle$ where $\mathcal{L}$ is a logical language and $\mathcal{R}$*

is a set of inference rules on this language, and a set of assumptions $\mathcal{A} \subseteq \mathcal{L}$, an argument for $c \in \mathcal{L}$ (the conclusion or claim) supported by $S \subseteq \mathcal{A}$ is a tree with nodes labelled by formulas in $\mathcal{L}$ or by the special symbol $\top$ such that:
- the root is labelled c
- for every node N
    - if N is a leaf then N is labelled either by an assumption (in S) or by $\top$
    - if N is not a leaf and b is the label of N, then there exists an inference rule $b \leftarrow b_1, \ldots, b_m$ ($m \geq 0$) and either $m = 0$ and the child of N is labelled by $\top$, or $m > 0$ and N has m children, labelled by $b_1, \ldots, b_m$ respectively
- S is the set of all assumptions labelling the leaves

We say that a set of assumptions $\mathcal{A}sms \subseteq \mathcal{A}$ enables the construction of an argument A (or alternatively, that A can be constructed based on $\mathcal{A}sms$) if A is supported by a subset of $\mathcal{A}sms$.

Notice that each assumption $\alpha \in \mathcal{A}$ enables an argument A_α with claim α supported by $\{\alpha\}$. That is, the corresponding tree has just one node that is labeled α.

Definition 2 ([12]). *An ABA framework is a tuple $\langle \mathcal{L}, \mathcal{R}, \mathcal{A}, \bar{} \rangle$ where:*
- $\langle \mathcal{L}, \mathcal{R} \rangle$ *is a deductive system*
- $\mathcal{A} \subseteq \mathcal{L}$ *is a (non-empty) set, whose elements are referred to as assumptions*
- $\bar{}$ *is a total mapping from $\mathcal{A}$ into $\mathcal{L}$, where $\overline{\alpha}$ is called the contrary of α*

For current purposes, we restrict ourselves to ABA-frameworks that are *flat* [4], meaning that no assumption is the head of an inference rule. Furthermore, we follow [12] in that each assumption has a unique contrary.

We are now ready to define the various abstract argumentation semantics (in the context of an ABA-framework). We say that an argument A_1 *attacks* an argument A_2 iff the conclusion of A_1 is the contrary of an assumption in A_2. Also, if $\mathcal{A}rgs$ is a set of arguments, then we write $\mathcal{A}rgs^+$ for $\{A \mid \text{there exists an argument in } \mathcal{A}rgs \text{ that attacks } A\}$. We say that a set of arguments $\mathcal{A}rgs$ is *conflict-free* iff $\mathcal{A}rgs \cap \mathcal{A}rgs^+ = \emptyset$. We say that a set of arguments $\mathcal{A}rgs$ *defends* an argument A iff each argument that attacks A is attacked by an argument in $\mathcal{A}rgs$.

Definition 3. *Let $\langle \mathcal{L}, \mathcal{R}, \mathcal{A}, \bar{} \rangle$ be an ABA framework, and let Ar be the set of arguments that can be constructed from this ABA framework. We say that $\mathcal{A}rgs \subseteq Ar$ is:*
- *a complete argument extension iff $\mathcal{A}rgs$ is conflict-free and $\mathcal{A}rgs = \{A \in Ar \mid \mathcal{A}rgs \text{ defends } A\}$*

- *a* grounded argument extension *iff it is the (subset-)minimal complete argument extension*

- *a* preferred argument extension *iff it is a (subset-)maximal complete argument extension*

- *a* semi-stable argument extension *iff it is a complete argument extension where $Args \cup Args^+$ is (subset-)maximal among all complete argument extensions*

- *a* stable argument extension *iff it is a complete argument extension where $Args \cup Args^+ = Ar$*

- *an* ideal argument extension *iff it is the (subset-)maximal complete argument extension that is contained in each preferred argument extension*

- *an* eager argument extension *iff it is the (subset-)maximal complete argument extension that is contained in each semi-stable argument extension*

It should be noticed that the grounded argument extension is unique, just like the ideal argument extension and the eager argument extension are unique [6]. Also, every stable argument extension is a semi-stable argument extension, and every semi-stable argument extension is a preferred argument extension [5]. Furthermore, if there exists at least one stable argument extension, then every semi-stable argument extension is a stable argument extension [5]. It also holds that the grounded argument extension is a subset of the ideal argument extension, which in its turn is a subset of the eager argument extension [6].

The next step is to describe the various ABA semantics. These are defined not in terms of sets of arguments (as is the case for abstract argumentation) but in terms of sets of assumptions. A set of assumptions $Asms_1$ is said to *attack* an assumption α iff $Asms_1$ enables the construction of an argument for conclusion $\overline{\alpha}$. A set of assumptions $Asms_1$ is said to attack a set of assumptions $Asms_2$ iff $Asms_1$ attacks some assumption $\alpha \in Asms_2$. Also, if $Asms$ is a set of assumptions, then we write $Asms^+$ for $\{\alpha \in \mathcal{A} \mid Asms \text{ attacks } \alpha\}$. We say that a set of assumptions $Asms$ is *conflict-free* iff $Asms \cap Asms^+ = \emptyset$. We say that a set of assumptions *defends* an assumption α iff each set of assumptions that attacks α is attacked by $Asms$.

Apart from the ABA-semantics defined in [10], we also define semi-stable and eager semantics in the context of ABA.[1]

Definition 4. *Let $\langle \mathcal{L}, \mathcal{R}, \mathcal{A}, \overline{} \rangle$ be an ABA framework, and let $Asms \subseteq \mathcal{A}$. We say that $Asms$ is:*

---

[1] Please notice that our definitions are slightly different from the ones in [10] (as we define all semantics in terms of complete assumption extensions) but equivalence is proved in the appendix.

- *a* complete assumption extension *iff* $Asms \cap Asms^+ = \emptyset$ *and* $Asms = \{\alpha \mid Asms$ *defends* $\alpha\}$

- *a* grounded assumption extension *iff it is the (subset-)minimal complete assumption extension*

- *a* preferred assumption extension *iff it is a (subset-)maximal complete assumption extension*

- *a* semi-stable assumption extension *iff it is a complete assumption extension where* $Asms \cup Asms^+$ *is (subset-)maximal among all complete assumption extensions*

- *a* stable assumption extension *iff it is a complete assumption extension where* $Asms \cup Asms^+ = \mathcal{A}$

- *an* ideal assumption extension *iff it is the (subset-)maximal complete assumption extension that is contained in each preferred assumption extension*

- *an* eager assumption extension *iff it is the (subset-)maximal complete assumption extension that is contained in each semi-stable assumption extension*

It should be noticed that the grounded assumption extension is unique, just like the ideal assumption extension and the eager assumption extension are unique. Also, every stable assumption extension is a semi-stable assumption extension, and every semi-stable assumption extension is a preferred assumption extension (Theorem 12 in the appendix, points 1 and 2). Furthermore, if there exists at least one stable assumption extension, then every semi-stable assumption extension is a stable assumption extension (Theorem 12 in the appendix, point 3). It also holds that the grounded assumption extension is a subset of the ideal assumption extension, which in its turn is a subset of the eager assumption extension (Theorem 16 in the appendix). Overall, we observe that in the context of ABA, semi-stable and eager semantics are well-defined and have properties that are similar to their abstract argumentation variants (as described in [5, 6]).[2]

---

[2]In essence, a semi-stable extension is a complete extension with maximal range (with *range* [22] being the union of the extension itself and what it attacks). These consist of arguments when applied to abstract argumentation, and of assumptions when applied to assumption-based argumentation. Although Definition 4 newly defines semi-stable semantics in the context of ABA, it does so in a very natural way following the already established concept of semi-stable semantics for abstract argumentation. Our interest in defining semi-stable semantics for ABA can further be justified by the recent research interest it has received [22, 5, 8, 2, 3, 1, 14] making it an important path to be explored.

3 Equivalence and Inequivalence

As can be observed from Definition 4 and Definition 3, the way assumption-based argumentation works is very similar to the way abstract argumentation works. In fact, there is a clear correspondence between these approaches, that allows one to convert ABA-extensions to abstract argumentation extensions, and vice versa. To formalise this correspondence we first define a function Asms2Args that maps assumptions extensions to argument extensions and a function Args2Asms that maps argument extensions to assumptions extensions.

Definition 5. *Let $\langle \mathcal{L}, \mathcal{R}, \mathcal{A}, \bar{} \rangle$ be an ABA framework, and let Ar be the set of all arguments that can be constructed using this ABA framework.*

- *We define $\text{Asms2Args} : 2^{\mathcal{A}} \to 2^{Ar}$ to be a function such that $\text{Asms2Args}(\mathcal{A}sms) = \{A \in Ar \mid A$ can be constructed based on $\mathcal{A}sms\}$*

- *We define $\text{Args2Asms} : 2^{Ar} \to 2^{\mathcal{A}}$ to be a function such that $\text{Args2Asms}(\mathcal{A}rgs) = \{\alpha \in \mathcal{A} \mid \alpha$ is an assumption occurring in an $A \in \mathcal{A}rgs\}$*

The next theorem shows that for certain semantics these functions indeed map assumption extensions to the corresponding arguments extensions and vice versa.

Theorem 6 ([10]). *Let $\langle \mathcal{L}, \mathcal{R}, \mathcal{A}, \bar{} \rangle$ be an ABA framework, and let Ar be the set of all arguments that can be constructed using this ABA framework.*

1. *If $\mathcal{A}sms \subseteq \mathcal{A}$ is a complete assumption extension, then $\text{Asms2Args}(\mathcal{A}sms)$ is a complete argument extension, and if $\mathcal{A}rgs \subseteq Ar$ is a complete argument extension, then $\text{Args2Asms}(\mathcal{A}rgs)$ is a complete assumption extension.*

2. *If $\mathcal{A}sms \subseteq \mathcal{A}$ is the grounded assumption extension, then $\text{Asms2Args}(\mathcal{A}sms)$ is the grounded argument extension, and if $\mathcal{A}rgs \subseteq Ar$ is the grounded argument extension, then $\text{Args2Asms}(\mathcal{A}rgs)$ is the grounded assumption extension.*

3. *If $\mathcal{A}sms \subseteq \mathcal{A}$ is a preferred assumption extension, then $\text{Asms2Args}(\mathcal{A}sms)$ is a preferred argument extension, and if $\mathcal{A}rgs \subseteq Ar$ is a preferred argument extension, then $\text{Args2Asms}(\mathcal{A}rgs)$ is a preferred assumption extension.*

4. *If $\mathcal{A}sms \subseteq \mathcal{A}$ is the ideal assumption extension, then $\text{Asms2Args}(\mathcal{A}sms)$ is the ideal argument extension, and if $\mathcal{A}rgs \subseteq Ar$ is the ideal argument extension, then $\text{Args2Asms}(\mathcal{A}rgs)$ is the ideal assumption extension.*

5. *If $\mathcal{A}sms \subseteq \mathcal{A}$ is a stable assumption extension, then $\text{Asms2Args}(\mathcal{A}sms)$ is a stable argument extension, and if $\mathcal{A}rgs \subseteq Ar$ is a stable argument extension, then $\text{Args2Asms}(\mathcal{A}rgs)$ is a stable assumption extension.*

Proof. Points 2 and 4 have been proved in [10], and point 5 has been proved in [21, Theorem 1],[3] so we only need to prove points 1 and 3. In the following, we slightly broaden terminology and say that argument A attacks assumption α iff the conclusion of A is $\overline{\alpha}$. Similarly, we say that a set of assumptions $Asms$ defends an argument A iff it defends each assumption in A, and we say that a set of arguments $Args$ defends an assumption α iff for each argument B with conclusion $\overline{\alpha}$, there is an argument $C \in Args$ that attacks B.

- **1, first conjunct:**
  Let $Asms \subseteq \mathcal{A}$ be a complete assumption extension and let $Args=\texttt{Asms2Args}(Asms)$. The fact that $Asms$ is conflict-free (that is $Asms \cap Asms^+ = \emptyset$) means one cannot construct an argument based on $Asms$ that attacks any assumption in $Asms$. Therefore, one cannot construct an argument based on $Asms$ that attacks any argument based on $Asms$. Hence, $Args$ is conflict-free (that is, $Args \cap Args^+ = \emptyset$).
  The fact that $Asms$ defends itself means that $Asms$ defends each assumption in $Asms$. Hence, $Asms$ defends each argument based on $Asms$ (each argument in $Args$). That is, $Args$ defends itself.
  The fact that each assumption defended by $Asms$ is in $Asms$ means that each argument whose assumptions are defended by $Asms$ is in $Args$. Hence, each argument defended by $Args$ is in $Args$.
  Altogether, we have observed that $Args$ is conflict-free and contains precisely the arguments it defends. That is, $Args$ is a complete argument extension.

- **1, second conjunct:**
  Let $Args \subseteq Ar$ be a complete argument extension and let $Asms = \texttt{Args2Asms}(Args)$. Suppose $Asms$ is not conflict-free. Then it is possible to construct an argument based on $Asms$ (say A) whose conclusion is the contrary of an assumption in $Asms$. A cannot be an element of $Args$ (otherwise $Args$ would not be conflict-free). From the thus obtained fact that $A \notin Args$, together with the fact that $Args$ is a complete argument extension, it follows that $Args$ does not defend A. But this is impossible, because $Args$ does defend all assumptions in A. Contradiction. Therefore, $Asms$ is conflict-free.
  The fact that $Args$ defends itself means that every $A \in Args$ is defended by $Args$, which implies that every assumption occurring in $Args$ is defended by $Args$, so every $\alpha \in Asms$ is defended by $Asms$. Hence, $Asms$ defends itself.
  The final thing to be shown is that $Asms$ contains every assumption it defends. Suppose $Asms$ defends $\alpha \in \mathcal{A}$. This means that for each argument B with conclusion $\overline{\alpha}$, $Asms$ enables the construction of an argument C that attacks B. The fact that all assumptions in C are found in arguments from $Args$ means that C is defended by

---

[3]Please note that our definition of ideal and stable semantics is slightly different than in [10, 21] but equivalence is proved in the appendix.

$Args$ (this is because $Args$ defends all its arguments). The fact that $Args$ is a complete argument extension then implies that $C \in Args$. This means that $Args$ defends the argument (say, A) consisting of the single assumption α. Hence, $A \in Args$, so $\alpha \in Asms$.

Altogether, we have observed that $Asms$ is conflict-free and contains precisely the assumptions it defends. That is, $Asms$ is a complete assumption extension.

- **3, first conjunct:**
  Let $Asms \subseteq \mathcal{A}$ be a preferred assumption extension and let $Args$=Asms2Args($Asms$). From point 1, it then follows that $Args$ is a complete assumption extension. Suppose, towards a contradiction, that $Args$ is not a *maximal* complete argument extension. Then there exists a complete argument extension $Args' \supsetneq Args$. Let $Asms' =$ Args2Asms($Args'$). It then follows that $Asms' \supseteq Asms$ and that $Asms' \neq Asms$,[4] so $Asms' \supsetneq Asms$. Moreover, from point 1 it follows that $Asms'$ is a complete assumption extension. But this would mean that $Asms$ is not a *maximal* complete assumption extension. Contradiction.

- **3, second conjunct:**
  Let $Args \subseteq Ar$ be a complete argument extension and let $Asms =$ Args2Asms($Args$). From point 1, it then follows that $Asms$ is a complete assumption extension. Suppose, towards a contradiction, that $Asms$ is not a *maximal* complete assumption extension. Then there exists a complete assumption extension $Asms' \supsetneq Asms$. Let $Args' =$ Asms2Args($Asms'$). It then holds that $Args' \supseteq Args$ and that $Args' \neq Args$,[5] so $Args' \supsetneq Args$. Moreover, from point 1 it follows that $Args'$ is a complete argument extension. But this would mean that $Args$ is not a *maximal* complete argument extension. Contradiction.

$\square$

Next we show that for complete semantics the functions Asms2Args and Args2Asms are also bijections.

---

[4]Suppose towards a contradiction that $Asms' = Asms$. From the fact that $Args' \supsetneq Args$ it follows that there exists an argument $A' \in Args'$ such that $A' \notin Args$. From the fact that $Asms' = Asms$ it follows that for each assumption $\alpha \in A'$ there exists an argument $A \in Args$ that contains α. Since each $A \in Args$ is defended by $Args$ (as $Args$ is a complete argument extension) it follows that A' is also defended by $Args$. From the fact that $Args$ is a complete argument extension, it then follows that $A \in Args$. Contradiction.

[5]The fact that $Asms' \supsetneq Asms$ implies there exists an assumption $\alpha' \in Asms'$ such that $\alpha' \notin Asms$. Let A' be the argument consisting of the single assumption α'. From the fact that $\alpha' \notin Asms$ it follows that $A' \notin Args$. However, from the definition of Asms2Args it does follow that $A' \in Args'$. Hence, $Args' \neq Args$.

Proposition 1. *When restricted to complete assumption extensions and complete argument extensions, the functions* Asms2Args *and* Args2Asms *become bijections and each other's inverses.*

Proof. Let $\mathcal{A}sms$ be a complete assumption extension and let $\mathcal{A}rgs$ be a complete argument extension. It suffices to prove statements (1) and (2) below.

1. Args2Asms(Asms2Args($\mathcal{A}sms$)) = $\mathcal{A}sms$

   (a) Suppose $\alpha \in \mathcal{A}sms$.
   Then by definition there exists an argument A_α consisting of a single assumption α. As $\alpha \in \mathcal{A}sms$ then also $A_\alpha \in$ Asms2Args($\mathcal{A}sms$). Therefore, $\alpha \in$ Args2Asms(Asms2Args($\mathcal{A}sms$)).

   (b) Suppose an $\alpha \in \mathcal{A}$ with $\alpha \notin \mathcal{A}sms$.
   Then there exists no argument in Asms2Args($\mathcal{A}sms$) that contains α. Therefore, $\alpha \notin$ Args2Asms (Asms2Args($\mathcal{A}sms$)).

2. Asms2Args(Args2Asms($\mathcal{A}rgs$)) = $\mathcal{A}rgs$.

   (a) Suppose $A \in \mathcal{A}rgs$.
   Then all assumptions used in A will be in Args2Asms($\mathcal{A}rgs$). This means that A can be constructed based on Args2Asms($\mathcal{A}rgs$). Therefore, $A \in$ Asms2Args (Args2Asms($\mathcal{A}rgs$)).

   (b) Suppose an $A \in Ar$ with $A \notin \mathcal{A}rgs$.
   The fact that $\mathcal{A}rgs$ is a complete argument extension implies that A is not defended by $\mathcal{A}rgs$. Therefore, there exists an argument $B \in Ar$ that attacks A, such that $\mathcal{A}rgs$ contains no C that attacks B. Assume, without loss of generality, that B attacks A by having a conclusion $\overline{\beta}$, where β is an assumption used in A. Then $\mathcal{A}rgs$ cannot contain any argument that uses assumption β (otherwise, this argument would not be defended against B, so $\mathcal{A}rgs$ would not be a complete arguments extension). Therefore, $\beta \notin$ Args2Asms($\mathcal{A}rgs$). This means that A cannot be constructed based on Args2Asms($\mathcal{A}rgs$). Therefore, $A \notin$ Asms2Args(Args2Asms($\mathcal{A}rgs$))

□

Now, as each preferred, grounded, stable, and ideal extension is also a complete extension, Proposition 1 also extends to these semantics. Combining Proposition 1 with Theorem 6 we obtain that under complete, grounded, preferred, stable and ideal semantics, argument extensions and assumption extensions are one-to-one related.

The above results might cause one to believe that similar observations can also be made for other semantics. Unfortunately, this is not always the case as we show next.

Theorem 7. *Let $\langle \mathcal{L}, \mathcal{R}, \mathcal{A}, \bar{} \rangle$ be an ABA framework, and let Ar be the set of all arguments that can be constructed using this ABA framework.*

1. *It is* not *the case that if $\mathcal{A}sms \subseteq \mathcal{A}$ is a semi-stable assumption extension, then $\mathtt{Asms2Args}(\mathcal{A}sms)$ is a semi-stable argument extension, and it is not the case that if $\mathcal{A}rgs \subseteq Ar$ is a semi-stable argument extension, then $\mathtt{Args2Asms}(\mathcal{A}rgs)$ is a semi-stable assumption extension.*

2. *It is* not *the case that if $\mathcal{A}sms \subseteq \mathcal{A}$ is an eager assumption extension, then $\mathtt{Asms2Args}(\mathcal{A}sms)$ is an eager argument extension, and it is not the case that if $\mathcal{A}rgs \subseteq Ar$ is an eager argument extension, then $\mathtt{Args2Asms}(\mathcal{A}rgs)$ is an eager assumption extension.*

Proof. Let $\mathcal{F}_{ex1} = \langle \mathcal{L}, \mathcal{R}, \mathcal{A}, \bar{} \rangle$ be an ABA framework with $\mathcal{L} = \{a, b, c, e, \alpha, \beta, \gamma, \epsilon\}$, $\mathcal{A} = \{\alpha, \beta, \gamma, \epsilon\}$, $\bar{\alpha} = a$, $\bar{\beta} = b$, $\bar{\gamma} = c$, $\bar{\epsilon} = e$ and $\mathcal{R} = \{r_1, r_2, r_3, r_4, r_5\}$ as follows:

$$r_1: c \leftarrow \gamma \qquad r_2: a \leftarrow \beta \qquad r_3: b \leftarrow \alpha \qquad r_4: c \leftarrow \gamma, \alpha \qquad r_5: e \leftarrow \epsilon, \beta$$

The following arguments can be constructed from this ABA framework.

- A_1, using the single rule r_1, with conclusion c and supported by $\{\gamma\}$
- A_2, using the single rule r_2, with conclusion a and supported by $\{\beta\}$
- A_3, using the single rule r_3, with conclusion b and supported by $\{\alpha\}$
- A_4, using the single rule r_4, with conclusion c and supported by $\{\gamma, \alpha\}$
- A_5, using the single rule r_5, with conclusion e and supported by $\{\epsilon, \beta\}$
- A_α, A_β, A_γ and A_ϵ, consisting of a single assumption α, β, γ and ϵ, respectively.

These arguments, as well as their attack relation, are shown in Figure 1. The complete argument extensions of AF_{ex1} are

$$\mathcal{A}rgs_1 = \emptyset, \ \mathcal{A}rgs_2 = \{A_2, A_\beta\}, \text{ and } \mathcal{A}rgs_3 = \{A_3, A_\alpha, A_\epsilon\}.$$

The associated complete assumption extensions of $\mathcal{F}_{ex1}$ are

$$\mathcal{A}sms_1 = \emptyset, \ \mathcal{A}sms_2 = \{\beta\}, \text{ and } \mathcal{A}sms_3 = \{\alpha, \epsilon\}.$$

Notice that, as one would expect, $\mathcal{A}rgs_i = \mathtt{Asms2Args}(\mathcal{A}sms_i)$ as well as $\mathcal{A}sms_i = \mathtt{Args2Asms}(\mathcal{A}rgs_i)$ for all $i \in \{1, 2, 3\}$.

It holds that $\mathcal{A}rgs_1 \cup \mathcal{A}rgs_1^+ = \emptyset$, $\mathcal{A}rgs_2 \cup \mathcal{A}rgs_2^+ = \{A_2, A_3, A_4, A_\alpha, A_\beta\}$ and $\mathcal{A}rgs_3 \cup \mathcal{A}rgs_3^+ = \{A_2, A_3, A_5, A_\alpha, A_\beta, A_\epsilon\}$, as well as $\mathcal{A}sms_1 \cup \mathcal{A}sms_1^+ = \emptyset$, $\mathcal{A}sms_2 \cup$

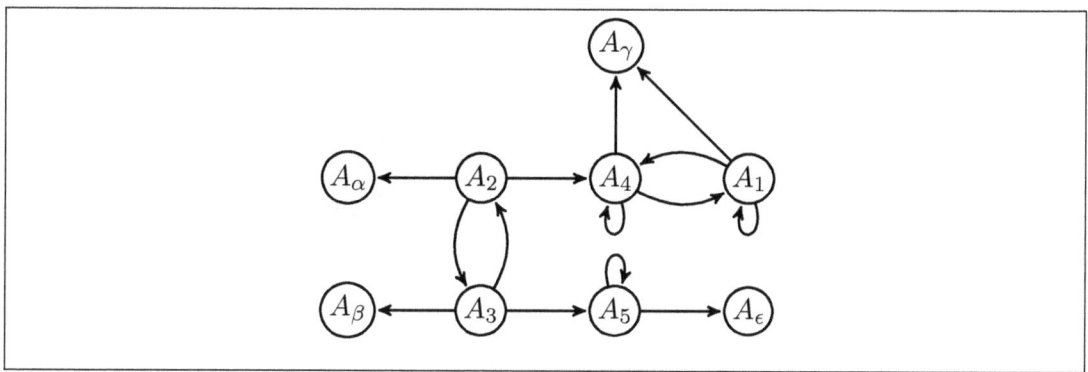

Figure 1: The argumentation framework AF_{ex1} associated with ABA framework $\mathcal{F}_{ex1}$.

$Asms_2^+ = \{\alpha, \beta\}$ and $Asms_3 \cup Asms_3^+ = \{\alpha, \beta, \epsilon\}$. Hence, $Args_2$ and $Args_3$ are semi-stable argument extensions, whereas only $Asms_3$ is a semi-stable assumption extension. We thus have a counterexample against the claim that if $Args$ ($Args_2$) is a semi-stable argument extension, $Asms = \texttt{Args2Asms}(Args)$ ($Asms_2$) is a semi-stable assumption extension.

We also observe that the eager argument extension is $Args_1$ whereas the eager assumption extension is $Asms_3$. Hence, we have a counterexample against the claim that if $Args$ is an eager argument extension then $Asms = \texttt{Args2Asms}(Args)$ is an eager assumption extension, as well as against the claim that is $Asms$ is an eager assumption extension then $Args = \texttt{Asms2Args}(Asms)$ is an eager argument extension.

The only thing left to be shown is that if $Asms$ is a semi-stable assumption extension, then $Args = \texttt{Asms2Args}(Asms)$ is not necessarily a semi-stable argument extension. For this, we slightly alter the ABA framework $\mathcal{F}_{ex1}$ by removing rule r_5 and the assumption ϵ (call the resulting ABA framework $\mathcal{F}_{ex2}$). Thus the arguments A_5 and A_ϵ no longer exists and hence $Args_3 = \{A_3, A_\alpha\}$. As now $Args_3 \cup Args_3^+ = \{A_2, A_3, A_\alpha, A_\beta\}$ is a proper subset of $Args_2 \cup Args_2^+$ the set $Args_3$ is no longer semi-stable. On the other side both $Asms_2 = \{\beta\}$, and $Asms_3 = \{\alpha\}$ are semi-stable assumption extensions. □

4 Discussion

The connection between assumption-based argumentation and abstract argumentation has received quite some attention in the literature. Dung *et al.*, for instance, claim that "ABA is an instance of abstract argumentation (AA), and consequently it inherits its various notions of 'acceptable' sets of arguments" [12]. Similarly, Toni claims that "ABA can be seen as an instance of AA, and (...) AA is an instance of ABA" [21]. While we agree that this holds for *some* of the admissibility-based semantics (like preferred and grounded), we have pointed out in the current paper that this certainly does not hold for *all* admissibility-based

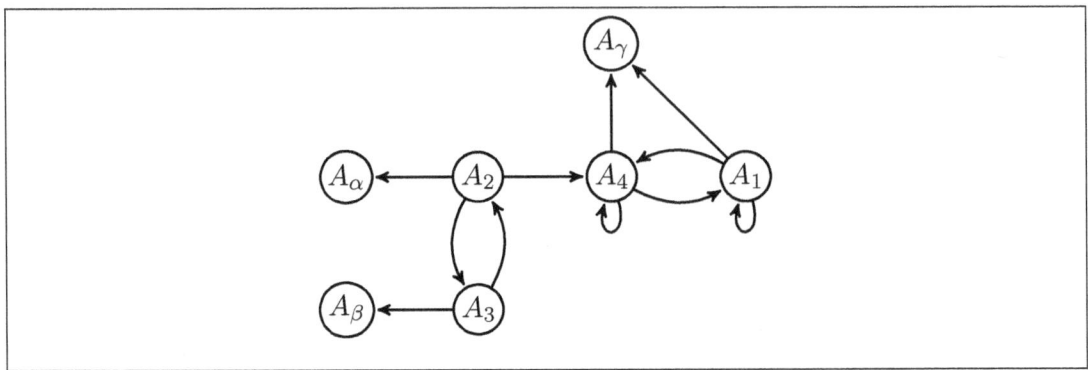

Figure 2: The argumentation framework AF_{ex2} associated with ABA framework $\mathcal{F}_{ex2}$.

semantics (semi-stable and eager). One could argue that claims like those above are perhaps a bit too general.

Prakken claims that "assumption-based argumentation (ABA) is a special case of the present framework [ASPIC+] with only strict inference rules, only assumption-type premises and no preferences." [18]. This claim is later repeated in the work of Modgil and Prakken, who state that "A well-known and established framework is that of assumption-based argumentation (ABA) [4], which (...) is shown (in [18])) to be a special case of the ASPIC+ framework in which arguments are built from assumption premises and strict inference rules only and in which all arguments are equally strong" [17]. However, we observe that the argumentation frameworks of Figure 1 and Figure 2 are counterexamples against this claim, in the context of semi-stable and eager semantics. These semantics, being admissibility-based, should work perfectly fine in the context of ASPIC+ (the rationality postulates of [7] would for instance be satisfied). Nevertheless, correspondence with ABA does not hold.

A possible criticism against our counter example of Figure 1 is that it uses a rule (r_4) that is subsumed by another rule (r_1). This raises the qeusion of whether counter examples still exist when no rule subsumes another rule. Our answer is affirmative: simply add an assumption δ and an atom d such that $\overline{\delta} = d$, replace r_1 by $c \leftarrow \gamma, \delta$ and add another rule (r_6) $d \leftarrow \delta$. For the resulting ABA theory, the semi-stable assumption extensions still do not correspond to the semi-stable argument extensions. Hence, the difference between ABA semi-stable (resp. ABA eager) and AA semi-stable (resp. AA eager) can be seen as a general phenomenon, that does not depend on whether some rules are subsumed by others.

Appendix: ABA semantics revisited

As mentioned earlier, the way the various ABA-semantics are defined in Definition 4 is slightly different from the way these were originally defined in [4, 10]. We have chosen to describe all ABA-semantics in a uniform way, based on the notion of complete semantics. This has been done not only for theoretical elegance, but also with an eye to practical applicability. For instance, it brings the advantage that once a particular property has been proven for each complete assumption extension, it has also been proven for each preferred, grounded, stable, semi-stable, ideal and eager assumption extension,[6] which can have benefits for future research. Furthermore, it paves the way for expressing ABA semantics in terms of assumption labellings, as is for instance done by Schulz and Toni, who define the concept of a complete assumption labelling [19]. In a similar way, one could then subsequently define the concept of a preferred assumption labelling as a complete assumption labelling where the set of in-labelled arguments is maximal, a grounded assumption labelling as a complete assumption labelling where the set of in-labelled arguments is minimal, etc. The fact that the extension-based ABA semantics have already been stated in terms of complete extensions (as is done in the current appendix) can then assist the task of proving equivalence between assumption-based extensions and assumption-based labellings.[7] The thus obtained theory of assumption labellings could then for instance be applied for examining the equivalence between ABA and logic programming (in the line of [19]), in a similar way as that the theory of argument labellings has been applied for examining the equivalence (and differences) between abstract argumentation and logic programming [9].

We will now proceed to show that our complete semantics based description of ABA-semantics in Definition 4 is equivalent to the original description of ABA-semantics in [4, 10]. We start with preferred semantics. Notice that a set of assumptions is called *admissible* iff it is conflict-free and defends each of its elements.

Theorem 8. *Let $\mathcal{F} = \langle \mathcal{L}, \mathcal{R}, \mathcal{A}, \bar{} \rangle$ be an ABA framework. The following two statements are equivalent:*

1. *$\mathcal{A}sms$ is a maximal admissible assumption set of $\mathcal{F}$*

2. *$\mathcal{A}sms$ is preferred assumption extension of $\mathcal{F}$*

Proof. From 1 to 2: Let $\mathcal{A}sms$ be a maximal admissible assumption set. It follows from [4, Corollary 5.8] that $\mathcal{A}sms$ is a complete assumption extension. Suppose $\mathcal{A}sms$ is not *maximal* complete. Then there exists a complete assumption extension $\mathcal{A}sms'$ with $\mathcal{A}sms \subsetneq \mathcal{A}sms'$. But since by definition, every complete assumption extension is also an admissible assumption set, it holds that $\mathcal{A}sms'$ is an admissible assumption set. But this would mean

---

[6]Similarly, in abstract argumentation, the fact that consistent outcome of argument-based entailment [7] has been proven under complete semantics implies that it has also been proven under preferred, grounded, stable, semi-stable, ideal and eager semantics.

[7]This would be in line with abstract argumentation, where reformulating some of the most common argument-based semantics in terms of complete semantics has been of great assistance for proving the equivalence between argument-based extensions and argument-based labellings.

that $Asms$ is not a *maximal* admissible assumption set. Contradiction.

From 2 to 1: Let $Asms$ be a maximal complete assumption extension. Then by definition, $Asms$ is also an admissible assumption set. We now need to prove that it is also a *maximal* admissible assumption set. Suppose this is not the case, then there exists a maximal admissible assumption set $Asms'$ with $Asms \subsetneq Asms'$. It follows from [4, Corollary 5.8] that $Asms'$ is also a complete assumption extension. But this would mean that $Asms$ is not a *maximal* complete assumption extension. Contradiction. □

The next thing to show is that our description of ideal semantics (Definition 4) coincides with that in [10]. More specifically, we will show that the notion of an ideal assumption extension is equivalent to that of a maximal ideal assumption set.

Definition 9. *Let $\mathcal{F} = \langle \mathcal{L}, \mathcal{R}, \mathcal{A}, \bar{} \rangle$ be an ABA framework. An* ideal assumption set *is defined as an admissible assumption set that is a subset of each preferred assumption extension.*

Lemma 1. *Let $\mathcal{F} = \langle \mathcal{L}, \mathcal{R}, \mathcal{A}, \bar{} \rangle$ be an ABA framework, and let $Asms_{id}$ be a maximal ideal assumption set. It holds that $Asms_{id}$ is a complete extension.*

Proof. Let $Asms_{id}$ be a maximal ideal assumption set. We only need to prove that if $Asms_{id}$ defends some $\alpha \in \mathcal{A}$ then $\alpha \in Asms_{id}$. Suppose $Asms_{id}$ defends α. Then every preferred assumption extension $Asms_p$ also defends α (this follows from $Asms_{id} \subseteq Asms_p$). As $Asms_p$ is also a complete extension, it follows that $\alpha \in Asms_p$. Hence, α is an element of every preferred assumption extension. Therefore, $Asms_{id} \cup \{\alpha\}$ is a subset of every preferred assumption extension. According to [4, Theorem 5.7], $Asms_{id} \cup \{\alpha\}$ is also an admissible set. From the fact that $Asms_{id}$ is a *maximal* ideal assumption set, and the trivial observation that $Asms_{id} \subseteq Asms_{id} \cup \{\alpha\}$, it then follows that $Asms_{id} = Asms_{id} \cup \{\alpha\}$. Therefore, $\alpha \in Asms_{id}$. □

Theorem 10. *Let $\mathcal{F} = \langle \mathcal{L}, \mathcal{R}, \mathcal{A}, \bar{} \rangle$ be an ABA framework and let $Asms \subseteq \mathcal{A}$. The following two statements are equivalent:*

1. *$Asms$ is a maximal ideal assumption set of $\mathcal{F}$*

2. *$Asms$ is an ideal assumption extension of $\mathcal{F}$ (in the sense of Definition 4)*

Proof. From 1 to 2: Let $Asms$ be a maximal ideal assumption set. It follows from Lemma 1 that $Asms$ is a complete assumption extension. Suppose $Asms$ is not a *maximal* complete assumption extension that is contained in every preferred assumption extension. Then there exists a complete assumption extension $Asms'$, with $Asms \subsetneq Asms'$, that is still contained in every preferred assumption extension. But since, by definition, every complete assumption extension is also an admissible assumption set, it holds that $Asms'$ is an admissible assumption set that is contained in every preferred assumption extension. That is, $Asms'$ is an ideal assumption set. But this would mean that $Asms$ is not a *maximal* admissible assumption set. Contradiction.

From 2 to 1: Let $Asms$ be an ideal assumption extension. Then, by definition, $Asms$ is also an ideal assumption set. We now need to prove that it is also a *maximal* ideal assumption set. Suppose this is not the case, then there exists a maximal ideal assumption set $Asms'$ with $Asms \subsetneq Asms'$. It follows from Lemma 1 that $Asms'$ is also a complete assumption extension. But this would mean that $Asms$ is not a *maximal* complete assumption extension that is contained in every preferred assumption extension. That is, $Asms$ is not an ideal assumption extension. Contradiction. $\square$

We proceed to show that our notion of stable semantics (Definition 4) coincides with the notion of stable semantics in [4].

Theorem 11. *Let $\mathcal{F} = \langle \mathcal{L}, \mathcal{R}, \mathcal{A}, \bar{} \rangle$ be an ABA framework, and let $Asms \subseteq \mathcal{A}$. The following two statements are equivalent:*

1. *$Asms$ does not attack itself and attacks each $\{\alpha\}$ with $\alpha \in \mathcal{A} \setminus Asms$*

2. *$Asms$ is a stable assumption extension of $\mathcal{F}$ (in the sense of Definition 4)*

Proof. From 1 to 2: Suppose $Asms$ does not attack itself and attacks each $\{\alpha\}$ with $\alpha \in \mathcal{A} \setminus Asms$. Then, according to [4, Theorem 5.5], $Asms$ is a complete extension. Moreover, the fact that $Asms$ attacks every $\{\alpha\}$ with $\alpha \in \mathcal{A} \setminus Asms$ means that $Asms \cup Asms^+ = \mathcal{A}$, so $Asms$ is a complete extension with $Asms \cup Asms^+ = \mathcal{A}$. That is, $Asms$ is a stable extension.

From 2 to 1: Suppose $Asms$ is a stable assumption extension. That is, $Asms$ is a complete assumption extension with $Asms \cup Asms^+ = \mathcal{A}$. From the fact that $Asms$ is a complete assumption extension, it follows that $Asms \cap Asms^+ = \emptyset$ so $Asms$ does not attack itself. From the fact $Asms \cup Asms^+ = \mathcal{A}$ it follows that $Asms^+ = \mathcal{A} \setminus Asms$, so $Asms$ attacks each $\{\alpha\}$ with $\alpha \in \mathcal{A} \setminus Asms$. $\square$

So far, we have examined our characterization of existing ABA-semantics (stable, preferred and ideal semantics) and found them to be equivalent to what have been stated in the literature. The next step is to focus on the ABA-semantics that have not yet been stated in the literature[8] (semi-stable and eager). Our aim is to show that, in the context of ABA, these semantics behave in a very similar way as they do in the context of abstract argumentation. We start with the relation between stable, semi-stable and preferred semantics.

Theorem 12. *Let $\mathcal{F} = \langle \mathcal{L}, \mathcal{R}, \mathcal{A}, \bar{} \rangle$ be an ABA framework. It holds that:*

---

[8] At least, not in the specific assumption-based ABA-context.

1. *every stable assumption extension is also a semi-stable assumption*

2. *every semi-stable assumption extension is also a preferred assumption extension*

3. *if there exists at least one stable assumption extension, then the stable assumption extensions and the semi-stable assumption extensions coincide*

Proof. 1. Let $Asms$ be a stable assumption extension of $\mathcal{F}$. Then, by definition, $Asms$ is a complete assumption extension with $Asms \cup Asms^+ = \mathcal{A}$. The fact that $Asms \cup Asms^+$ is $\mathcal{A}$ implies that it is maximal (by definition, it cannot be a proper superset of $\mathcal{A}$). Hence, $Asms$ is a complete assumption extension where $Asms \cup Asms^+$ is maximal. That is, $Asms$ is a semi-stable assumption extension.

2. Let $Asms$ be a semi-stable assumption extension of $\mathcal{F}$. Then, by definition, $Asms$ is a complete assumption extension where $Asms \cup Asms^+$ is maximal. We now show that $Asms$ itself is also maximal. Suppose there is a complete assumption extension $Asms'$ with $Asms \subsetneq Asms'$. Then, from the fact that the $^+$-operator is monotonic, it follows that $Asms^+ \subseteq Asms'^+$. This, together with the fact that $Asms \subsetneq Asms'$ implies that $Asms \cup Asms^+ \subsetneq Asms' \cup Asms'^+$. But that would mean that $Asms$ is not a semi-stable assumption extension. Contradiction. Therefore, $Asms$ is a maximal complete assumption extension. That is, $Asms$ is a preferred assumption extension.

3. Suppose there exists at least one stable assumption extension ($Asms_{st}$). The fact that every stable assumption extension is also a semi-stable assumption extension has already been proven by point 1, so the only thing left to prove is that every semi-stable assumption extension is also a stable assumption extension. Let $Asms$ be a semi-stable assumption extension. Then, by definition, $Asms$ is a complete assumption extension where $Asms \cup Asms^+$ is maximal. From the fact that $Asms_{st}$ is a complete assumption extension with $Asms_{st} \cup Asms_{st}^+ = \mathcal{A}$, it follows that for $Asms \cup Asms^+$ to be maximal, it has to be $\mathcal{A}$ as well. This implies that $Asms$ is a stable assumption extension. □

We proceed to examine the concept of eager semantics in the context of ABA. Our aim is to show that the eager assumption extension is unique. In order to do so, we first need to define the concept of an eager assumption set. Notice that an eager assumption set relates to the eager assumption extension in the same way as an ideal assumption set relates to the ideal assumption extension.

Definition 13. Let $\mathcal{F} = \langle \mathcal{L}, \mathcal{R}, \mathcal{A}, \bar{} \rangle$ be an ABA framework. An eager assumption set is defined as an admissible assumption set that is a subset of each semi-stable assumption extension.

Theorem 14. Let $\mathcal{F} = \langle \mathcal{L}, \mathcal{R}, \mathcal{A}, \bar{} \rangle$ be an ABA framework. There exists precisely one maximal eager assumption set.

Proof. We first prove that there exists at least one maximal eager assumption set. This is relatively straightforward, because there exists at least one eager assumption set (the empty set), which together with the fact that that there are only finitely many eager assumption sets (which follows from the fact that $\mathcal{A}$ is finite) implies that there exists at least one maximal eager assumption set.

The next thing to prove is that there exists at most one maximal eager assumption set. Let $Asms_1$ and $Asms_2$ be maximal eager assumption sets. From the fact that for each semi-stable assumption extension $Asms_{sem}$, it holds that $Asms_1 \subseteq Asms_{sem}$ and $Asms_2 \subseteq Asms_{sem}$ it follows that $Asms_1$ and $Asms_2$ do not attack each other (otherwise $Asms_{sem}$ would attack itself). Hence, $Asms_3 = Asms_1 \cup Asms_2$ does not attack itself. Also, $Asms_3$ defends itself, as $Asms_1$ and $Asms_2$ defend themselves. Hence, $Asms_3$ is an admissible assumption set that is a subset of each semi-stable assumption extension. That is, $Args_3$ is an eager assumption set. Also, from the fact that $Asms_3 = Asms_1 \cup Asms_2$, it follows that $Asms_1 \subseteq Asms_3$ and $Asms_2 \subseteq Asms_3$. From the fact that $Asms_1$ and $Asms_2$ are *maximal* eager assumption sets, it then follows that $Asms_1 = Asms_3$ and $Asms_2 = Asms_3$. Therefore, $Asms_1 = Asms_2$. □

Lemma 2. Let $\mathcal{F} = \langle \mathcal{L}, \mathcal{R}, \mathcal{A}, \bar{} \rangle$ be an ABA framework, and let $Asms_{eag}$ be the maximal eager assumption set. It holds that $Asms$ is a complete assumption extension.

Proof. Let $Asms_{eag}$ be a maximal eager assumption set. We only need to prove that if $Asms_{eag}$ defends some $\alpha \in \mathcal{A}$ then $\alpha \in Asms_{eag}$. Suppose $Asms_{eag}$ defends α. Then every semi-stable assumption extension $Asms_{sem}$ also defends α (this follows from $Asms_{eag} \subseteq Asms_{sem}$). As $Asms_{sem}$ is also a complete assumption extension, it follows that $\alpha \in Asms_{sem}$. Hence, α is an element of every semi-stable assumption extension. Therefore, $Asms_{eag} \cup \{\alpha\}$ is a subset of every semi-stable assumption extension. According to [4, Theorem 5.7], $Asms_{eag} \cup \{\alpha\}$ is also an admissible assumption set. Hence, $Asms_{eag} \cup \{\alpha\}$ is an eager assumption set. From the fact that $Asms_{eag}$ is a *maximal* eager assumption set, and the trivial observation that $Asms_{eag} \subseteq Asms_{eag} \cup \{\alpha\}$, it then follows that $Asms_{eag} = Asms_{eag} \cup \{\alpha\}$. Therefore, $\alpha \in Asms_{eag}$. □

Theorem 15. Let $\mathcal{F} = \langle \mathcal{L}, \mathcal{R}, \mathcal{A}, \bar{} \rangle$ be an ABA framework and let $Asms \subseteq \mathcal{A}$. The following two statements are equivalent:

1. $Asms$ is a maximal eager assumption set of $\mathcal{F}$

2. $\mathcal{A}sms$ is an eager assumption extension of $\mathcal{F}$ (in the sense of Definition 4)

Proof. From 1 to 2: Let $\mathcal{A}sms$ be a maximal eager assumption set. It follows from Lemma 2 that $\mathcal{A}sms$ is a complete assumption extension. Suppose $\mathcal{A}sms$ is not a *maximal* complete assumption extension that is contained in every semi-stable assumption extension. Then there exists a complete assumption extension $\mathcal{A}sms'$, with $\mathcal{A}sms \subsetneq \mathcal{A}sms'$, that is still contained in every semi-stable assumption extension. But since by definition, every complete assumption extension is also an admissible assumption set, it holds that $\mathcal{A}sms'$ is an admissible assumption set that is contained in every semi-stable assumption extension. That is, $\mathcal{A}sms'$ is an eager assumption set. But this would mean that $\mathcal{A}sms$ is not a *maximal* eager assumption set. Contradiction.

From 2 to 1: Let $\mathcal{A}sms$ be an eager assumption extension. Then, by definition, $\mathcal{A}sms$ is also an eager assumption set. We now need to prove that it is also a *maximal* eager assumption set. Suppose this is not the case, then there exists a maximal eager assumption set $\mathcal{A}sms'$ with $\mathcal{A}sms \subsetneq \mathcal{A}sms'$. It follows from Lemma 2 that $\mathcal{A}sms'$ is also a complete assumption extension. But this would mean that $\mathcal{A}sms$ is not a *maximal* complete assumption extension that is contained in every semi-stable assumption extension. That is, $\mathcal{A}sms$ is not an eager assumption extension. Contradiction. □

Theorem 16. *Let $\mathcal{F} = \langle \mathcal{L}, \mathcal{R}, \mathcal{A}, \bar{} \rangle$ be an ABA framework. It holds that:*

1. *the ideal assumption extension is a superset of the grounded assumption extension*

2. *the eager assumption extension is a superset of the ideal assumption extension*

Proof. 1. This follows from the fact that the ideal assumption extension is a complete assumption extension, together with the fact that the grounded assumption extension is the unique minimal (w.r.t. set inclusion) complete assumption extension.

2. This follows from the fact that every semi-stable assumption extension is also a preferred assumption extension, together with the definitions of the ideal (resp. eager) assumption extensions as the maximal complete assumption extension that is a subset of every preferred (resp. semi-stable) assumption extension.

□

Acknowledgements

The first author has been supported by the National Research Fund, Luxembourg (LAAMI project) and by the Engineering and Physical Sciences Research Council (EPSRC, UK), grant ref. EP/J012084/1 (SAsSy project). The second and third authors have been supported by CNPq (Universal 2012 - Proc. no. 473110/2012-1), CAPES (PROCAD 2009) and CNPq/CAPES (Casadinho/PROCAD 2011).

References

[1] P. Baroni, M.W.A. Caminada, and M. Giacomin. An introduction to argumentation semantics. *Knowledge Engineering Review*, 26(4):365–410, 2011.

[2] P. Baroni and M. Giacomin. On principle-based evaluation of extension-based argumentation semantics. *Artificial Intelligence*, 171(10-15):675–700, 2007.

[3] P. Baroni and M. Giacomin. Skepticism relations for comparing argumentation semantics. *Int. J. Approx. Reasoning*, 50(6):854–866, 2009.

[4] A. Bondarenko, P.M. Dung, R.A. Kowalski, and F. Toni. An abstract, argumentation-theoretic approach to default reasoning. *Artificial Intelligence*, 93:63–101, 1997.

[5] M.W.A. Caminada. Semi-stable semantics. In P.E. Dunne and T.J.M. Bench-Capon, editors, *Computational Models of Argument; Proceedings of COMMA 2006*, pages 121–130. IOS Press, 2006.

[6] M.W.A. Caminada. Comparing two unique extension semantics for formal argumentation: ideal and eager. In Mohammad Mehdi Dastani and Edwin de Jong, editors, *Proceedings of the 19th Belgian-Dutch Conference on Artificial Intelligence (BNAIC 2007)*, pages 81–87, 2007.

[7] M.W.A. Caminada and L. Amgoud. On the evaluation of argumentation formalisms. *Artificial Intelligence*, 171(5-6):286–310, 2007.

[8] M.W.A. Caminada, W.A. Carnielli, and P.E. Dunne. Semi-stable semantics. *Journal of Logic and Computation*, 22(5):1207–1254, 2012.

[9] M.W.A. Caminada, S. Sá, and J. Alcântara. On the equivalence between logic programming semantics and argumentation semantics. In Linda C. van der Gaag, editor, *Symbolic and Quantitative Approaches to Reasoning with Uncertainty; Proceedings of ECSQARU 2013*, pages 97–108, 2013.

[10] P. M. Dung, P. Mancarella, and F. Toni. Computing ideal sceptical argumentation. *Artificial Intelligence*, 171(10-15):642–674, 2007.

[11] P.M. Dung. On the acceptability of arguments and its fundamental role in nonmonotonic reasoning, logic programming and n-person games. *Artificial Intelligence*, 77:321–357, 1995.

[12] P.M. Dung, R.A. Kowalski, and F. Toni. Assumption-based argumentation. In Guillermo Simari and Iyad Rahwan, editors, *Argumentation in Artificial Intelligence*, pages 199–218. Springer US, 2009.

[13] W. Dvořák, P.E. Dunne, and S. Woltran. Parametric properties of ideal semantics. In *Proceedings IJCAI 2011*, pages 851–856, 2011.

[14] W. Dvořák and S. Woltran. Complexity of semi-stable and stage semantics in argumentation frameworks. *Information Processing Letters*, 110:425–430, 2010.

[15] A.J. García, J. Dix, and G.R. Simari. Argument-based logic programming. In Guillermo Simari and Iyad Rahwan, editors, *Argumentation in Artificial Intelligence*, pages 153–171. Springer US, 2009.

[16] G. Governatori, M.J. Maher, G. Antoniou, and D. Billington. Argumentation semantics for defeasible logic. *Journal of Logic and Computation*, 14(5):675–702, 2004.

[17] S.J. Modgil and H. Prakken. A general account of argumentation with preferences. *Artificial*

Intellligence, 195:361–397, 2013.

[18] H. Prakken. An abstract framework for argumentation with structured arguments. *Argument and Computation*, 1(2):93–124, 2010.

[19] C. Schulz and F. Toni. Logic programming in assumption-based argumentation revisited — semantics and graphical representation. In *Proceedings of AAAI 2015*, 2015. in print.

[20] G.R. Simari and R.P. Loui. A mathematical treatment of defeasible reasoning and its implementation. *Artificial Intelligence*, 53:125–157, 1992.

[21] F. Toni. Reasoning on the web with assumption-based argumentation. In Thomas Eiter and Thomas Krennwallner, editors, *Reasoning Web. Semantic Technologies for Advanced Query Answering*, pages 370–386, 2012.

[22] B. Verheij. Two approaches to dialectical argumentation: admissible sets and argumentation stages. In J.-J.Ch. Meyer and L.C. van der Gaag, editors, *Proceedings of the Eighth Dutch Conference on Artificial Intelligence (NAIC'96)*, pages 357–368, 1996.

[23] G.A.W. Vreeswijk. Abstract argumentation systems. *Artificial Intelligence*, 90:225–279, 1997.

[24] Y. Wu, M.W.A. Caminada, and D.M. Gabbay. Complete extensions in argumentation coincide with 3-valued stable models in logic programming. *Studia Logica*, 93(1-2):383–403, 2009. Special issue: new ideas in argumentation theory.

Physarum Polycephalum Diagrams for Syllogistic Systems

Andrew Schumann
Department of Cognitivistics, University of Information Technology and Management, Rzeszow, Poland
andrew.schumann@gmail.com

Andrew Adamatzky
Unconventional Computing Centre, UWE, Bristol, UK
andrew.adamatzky@uwe.ac.uk

Abstract

In this paper, we propose syllogistic circuits on the medium of *Physarum Polycephalum* plasmodium. These circuits are designed by chemical signals stimulating the plasmodium propagation. As a result, the circuits are built up on continuously growing plasmodia and any their motions are considered syllogistic conclusions. We show how we can implement different syllogistic systems on the medium of *Physarum Polycephalum* plasmodia within these circuits, namely we implement the following systems: the Aristotelian syllogistic, the performative syllogistic, and Talmudic reasoning by *qal wa-homer*. In this way, we can design biological devices, where inputs are represented not by electrical signals, but by chemical signals in accordance with the plasmodium chemotaxis, and outputs are represented by the plasmodium behaviour.

Keywords: Syllogisms

1 Introduction

One of the first logicians who proposed a spatial implementation of Aristotelian syllogistic reasoning was Lewis Carroll [3], [4]. He used three kinds of syllogistic propositions: (i) the universal affirmative ('all members of its subject are members of its predicate'), (ii) the universal negative ('no members of its subject are members

---

This research is supported by FP7-ICT-2011-8.

of its predicate'), (iii) and the particular affirmative ('some members of its subject are members of its predicate'). His examples are as follows: 'All red apples are ripe', 'No red apples are ripe', 'Some red apples are ripe'. For verifying syllogistic propositions he proposed the following biliteral diagram: $\begin{array}{|c|c|} \hline xy & xy' \\ \hline x'y & x'y' \\ \hline \end{array}$ that plays the role of 'universe of discourse' for all syllogistic propositions over adjuncts x, y, non-x (which is denoted by x'), non-y (which is denoted by y'). For example, let x mean 'old,' so that x' will mean 'new'. Let y mean 'English,' so that y' will mean 'foreign'. Assume that 'books' are an appropriate universe of discourse. Then we can divide this universe into the following four classes: xy ('old English books'), xy' ('old foreign books'), $x'y$ ('new English books'), $x'y'$ ('new foreign books').

Now let us take two kinds of counters: grey and black. If a black counter is placed within a cell, this means that "this cell is occupied" (i.e. "there is at least one thing in it"). If a grey counter is placed within a cell, this means that "this cell is empty" (i.e. "there is nothing in it"). Thus, using grey and black counters we can verify all the basic syllogistic propositions.

In this paper, we develop Carroll's ideas, but our diagrams will be represented by living organisms, namely by Physarum behaviours. In *Physarum Chip Project: Growing Computers From Slime Mould* supported by FP7 we are going to build a programmable amorphous biological computer. In this computer, logic circuits are represented by programmable behaviours of *Physarum Polycephalum* plasmodium – the one-call organism that behaves according to different stimuli called attractants and repellents and propagates networks connecting all reachable food sources. When designing an object-oriented programming language for the simulation of plasmodium behaviours [16], we detect a possibility on the media of *Physarum* to construct logic circuits for different symbolic-logical systems, including syllogistic systems. The circuits for various syllogistics may be built in the way of Carroll's diagrams. One of the most interesting outcomes of this research is that, first, the universe of discourse is considered a continuously growing living organism (*Physarum* plasmodium), second, any motions of this organism are considered as inferring syllogistic conclusions. Carroll's like diagrams allow us to define all possible directions in any motion of *Physarum*.

In this paper, we show how we can implement the Aristotelian syllogistic (section 4), the performative syllogistic (section 5), and Talmudic reasoning (section 6) via *Physarum* plasmodium behaviour.

2 Biological Computations by Chemical Signals

The plasmodium propagation can be interpreted as a kind of computation which can solve different tasks which assume concurrency: maze-solving [9], solving the Steiner Problem [20], minimum-risk path finding [10], [11], solving the traffic optimization problem [24], associative learning [18], memorizing and anticipating repeated events [12], etc. The feature of these computations is that a plasmodium black box always has much more outputs than inputs. We can generalize this property and claim that this feature concerns any computation on behavioural systems of living organisms. It is related to the problem of free will, where living organisms face a choice among several outputs given just one input: they can choose just one or simultaneously many outputs according to their will. Evidently, this is not typical for the electronic devices.

Hence, we can assume that it is possible to design a biological device, where instead of electrical signals the calculation process is performed by using the plasmodium *chemotaxis* [2] (about other approaches to information processing by chemotaxis see [6]), i.e. knowledge of the typical reactions of plasmodium to chemical signals attracting or repelling the plasmodium behaviour. In this device, the number of outputs usually is much larger than the number of inputs, where inputs are chemical signals attracting or repelling the behaviour and outputs are behavioural responses. We can artificially delete some outputs by using additional repellents. Nevertheless, as more repellents in complex gates as lower accuracy of computations in this device.

Notably, the accuracy of implementing classical logics and Turing-complete machines, such as Kolmogorov-Uspensky machines [22], [1], in the plasmodium behaviour is low. For instance, let computing circuits be implemented using a collision-based computing approach, where the plasmodium propagates on non-nutrient substrate in the form of a compact wave-fragment of protoplasm. Logic gates can be constructed in the laboratory conditions as follows: $\langle x, y \rangle \to \langle x \text{ AND } y, x \text{ OR } y \rangle$ and $\langle x, y \rangle \to \langle x, \text{ NOT } x \text{ AND } y \rangle$. Then the accuracy of experimental laboratory prototypes of the gate $\langle x, y \rangle \to \langle x \text{ AND } y, x \text{ OR } y \rangle$ was over 69% and of the gate $\langle x, y \rangle \to \langle x, \text{ NOT } x \text{ AND } y \rangle$ was over 59%. In the frequency-based Boolean logical gates implemented with *Physarum* based on frequencies of oscillations, the accuracy is as follows: 90% for OR/NOR, 77.8% for AND/NAND, 91.7% for NOT, 70.8% for XOR/XNOR. Notice that as more complex circuits, as lower accuracy of their implementation.

This problem with accuracy is linked to the fact that the plasmodium as living organism wants to be propagated in all possible directions (to possess more outputs than inputs). This means that the plasmodium does not follow the induction principle to satisfy the minimal conditions in funding and occupying the food sources, e.g. by shortest distances. In other words, the plasmodium behaviour is context-dependent and it is not a kind of spatial Turing-complete machines such as Kolmogorov-Uspensky machines. We can implement these machines only with some accuracy using repellents, because to follow these machines and classical logic gates is not natural for behavioural systems.

Usually, computations on *Physarum* are studied to approximate transportation systems and hierarchies of planar proximity graphs, e.g. to approximate concurrent phenomena. In our research, we are going to obtain an abstract chip based on plasmodium and we need to find out logical systems which are natural for the plasmodium behaviour and, therefore, they could be implemented with a high accuracy.

In this paper, we propose the implementation of Aristotelian syllogistic of [8] as a system whose semantics can be expressed by Kolmogorov-Uspensky machines. In this implementation we need many repellents and, thus, the behaviour of plasmodium is not so natural here. Then we show how we can implement the performative syllogistic of [15]. For this implementation we do not need repellents at all. The performative syllogistic, on the one hand, is more expressive than the Aristotelian syllogistic (it contains all tautologies of the latter, see [15]), on the other hand, its semantics is context-based and depends on neighbors for plasmodium active zones. Due to this semantics, the performative syllogistic is sound and complete on plasmodium propagations which are performed without repellents. This semantics cannot be expressed by Kolmogorov-Uspensky machines.

Logic circuits constructed on the basis of the performative syllogistic seem to be natural for behavioural systems and these circuits have very high accuracy in implementing. Our general motivation in designing logic circuits in behavioural systems without repellents is as follows:

- in this way, we can present behavioural systems as a calculation process more naturally;

- we can design devices, where there are much more outputs than inputs, for performing massive-parallel computations in the bio-inspired way;

- we can obtain unconventional (co)algorithms by programming behavioural systems.

Biological and bio-inspired computations allow us to perform calculations through behaviours of living organisms. The main disadvantage of these computations is that biological devices are much slower than digital computers, but they can effectively solve complex massive-parallel tasks.

3 Strings in the *Physarum* Growing Universe

Physarum Polycephalum behaves by plasmodia which can have the form either waves or protoplasmic tubes. Plasmodia grow from active zones concurrently. At these active zones, three basic operations stimulated by nutrients (attractants) and some other conditions can be observed: fusion, multiplication, direction, and repelling operations. (i) The *fusion*, $Fuse$, means that two active zones A_1 and A_2 produce a new active zone A_3. (ii) The *multiplication*, $Mult$, means that the active zone A_1 splits into two independent active zones A_2 and A_3 propagating along their own trajectories. (iii) The *direction*, $Direct$, means that the active zone A moves to a source of nutrients. (iv) The *repelling*, $Repel$, means that the active zone A avoids repellents. These operations, $Fuse$, $Mult$, $Direct$, $Repel$ can be determined by the following stimuli: (1) the set of attractants $\{N_1, N_2, \dots\}$, sources of nutrients, on which the plasmodium feeds; (2) the set of repellents $\{R_1, R_2, \dots\}$, light and some thermo- and salt-based conditions.

The universe, where *Physarum* lives, consists of cells possessing different topological properties according to the intensity of chemo-attractants and chemo-repellents. The intensity entails the natural or geographical neighborhood of the set's elements in accordance with the spreading of attractants or repellents. As a result, we obtain Voronoi cells. Let us define what they are mathematically. Let P be a nonempty finite set of planar points and $|P| = n$. For points $p = (p_1, p_2)$ and $x = (x_1, x_2)$ let $d(p, x) = \sqrt{(p_1 - x_1)^2 + (p_2 - x_2)^2}$ denote their Euclidean distance. A planar Voronoi diagram of the set P is a partition of the plane into cells, such that for any element of P, a cell corresponding to a unique point p contains all those points of the plane which are closer to p in respect to the distance d than to any other node of P. A unique region

$$vor(p) = \bigcap_{m \in P, m \neq p} \{z \in \mathbf{R}^2 : d(p, z) < d(m, z)\}$$

assigned to the point p is called the *Voronoi cell* of the point p. Within one Voronoi cell a reagent has the full power to attract or repel the plasmodium. The distance d is defined by intensity of reagent spreading like in other chemical reactions simulated by Voronoi diagrams. When two spreading wave fronts of two reagents meet, this means that the plasmodium cannot choose any further direction, and splits (see fig.1). Within the same Voronoi cell two active zones will fuse.

If a Voronoi center is presented by an attractant a that is activated and occupied by the plasmodium, this means that there exists a string a. This string has the meaning "a exists". If a Voronoi center is presented by a repellent $[a]$ that is activated and avoided by the plasmodium, this means that there exists a string $[a]$. This string

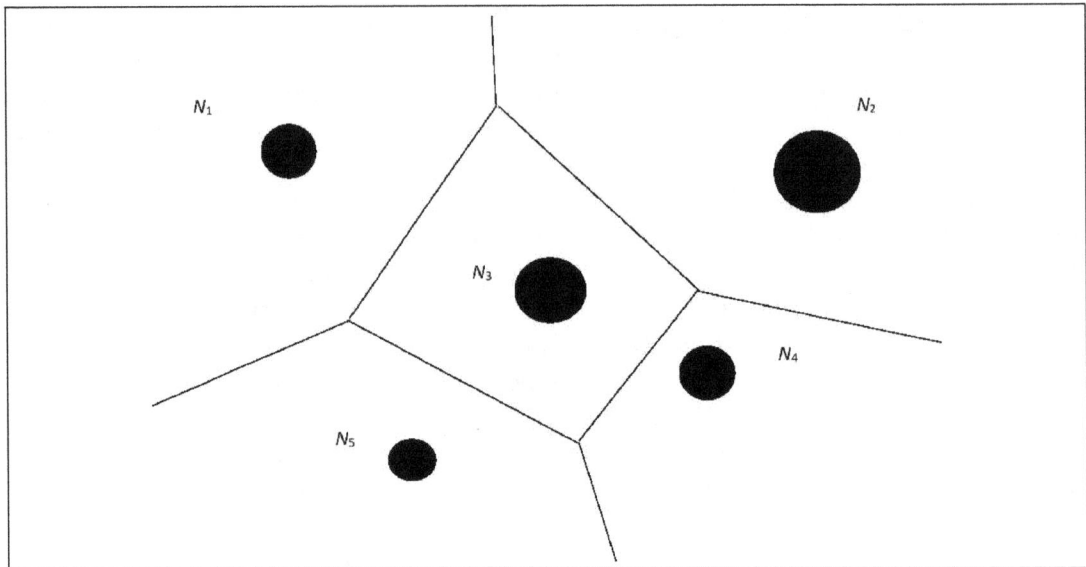

Figure 1: The Voronoi diagram for *Physarum*, where different attractants have different intensity and power.

has the meaning "a does not exist". If two neighbor Voronoi cells contain activated attractants a and b, which are occupied by the plasmodium, and between both centers there are protoplasmic tubes, then we say that there exists a string ab and a string ba. The meaning of those strings is equal and it is as follows: "ab exist", "ba exist", "some a is b", "some b is a".

If one neighbor Voronoi cell contains an activated attractant a which is occupied by the plasmodium and another neighbor Voronoi cell contains an activated repellent $[b]$ which is avoided by the plasmodium, then we say that there exists a string $a[b]$ and a string $[b]a$. The meaning of those strings is equal and it is as follows: "ab do not exist, but a exists without b", "there exists a and no a is b", "no b is a and there exists a", "a exists and b does not exist".

If two neighbor Voronoi cells contain activated repellents $[a]$ and $[b]$ which are avoided by the plasmodium, then there exists a string $[ab]$ and a string $[ba]$. The meaning of those strings is equal and it is as follows: "ab do not exist together", "there are no a and there are no b", "no b is a", "no a is b".

Thus, the family of strings between nearest attractants presents a proximity graph which continuously grows from one attractant to another. This expansion could be demonstrated as a *Toussaint hierarchy* [1], where the family of strings starts from a nearest-neighborhood graph and any next graph in the hierarchy is

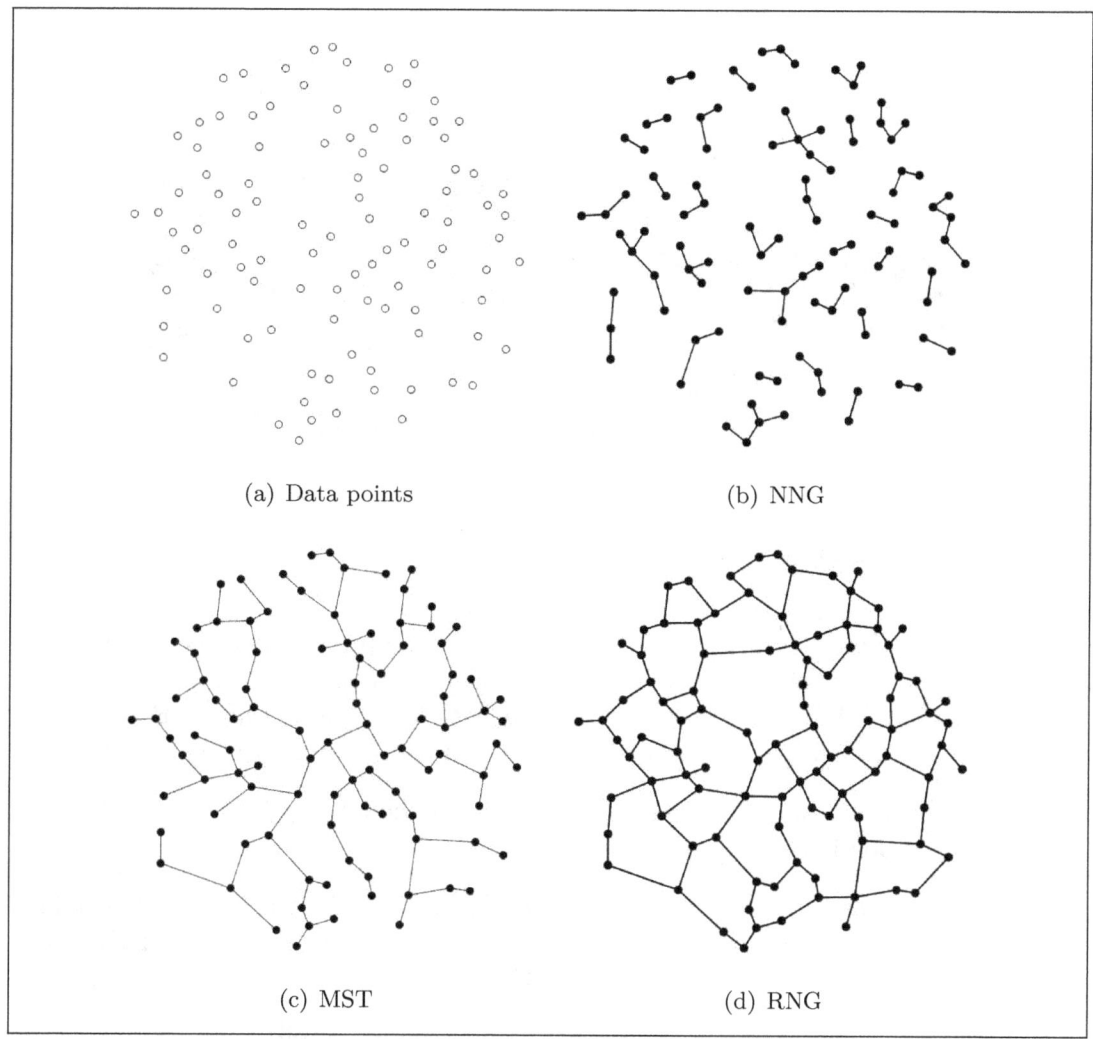

Figure 2: Examples of proximity graphs (part 1).

produced from previous graph by adding some edges between non-adjacent nodes, see fig.2:

$$\text{NNG} \to \text{MST} \to \text{RNG} \to \text{GG} \to \text{DT},$$

where

1. NNG is a *nearest-neighborhood graph*. It is the simplest and possibly most

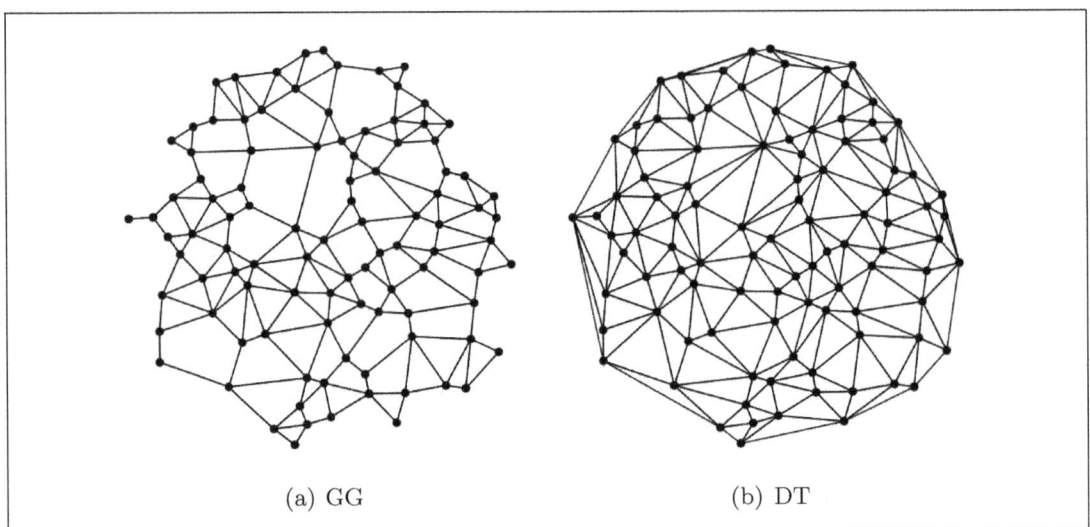

Figure 3: Examples of proximity graphs (part 2).

natural of proximity graphs. A point in the graph is connected by an edge to its nearest neighbor. Given planar set $\mathbf{V}$ we can define the graph as follows: $\mathrm{NNG}(\mathbf{V}) = \langle \mathbf{V}, \mathbf{E} \rangle$, where for $a, b \in \mathbf{V}$ we have $(ab) \in \mathbf{E}$ iff $|ab| = \min_{c \in \mathbf{V} - \{\mathbf{a}\}} |ac|$. In general case NNG is a disconnected directed graph.

2. MST is a *minimal spanning tree*. It is a connected acyclic graph which has minimal possible sum of edges' lengths.

3. RNG is a *relative neighborhood graph*. It is a graph, where any two points (a, b) are connected by an edge if the intersection of open disks of radius $|ab|$ centered at a and b is empty: $(ab) \in \mathbf{E}$ iff $|ab| \leq \max_{c \in \mathbf{V} - \{a,b\}} \{|ac|, |bc|\}$.

4. GG is a *Gabriel graph*. It is a graph, where points a and b are connected by an edge if the closed disk having the segment (ab) as its diameter is empty: $(ab) \in \mathbf{E}$ iff $|ab| \geq \min_{c \in \mathbf{V} - \{a,b\}} \{|\frac{a+b}{2} c|\}$.

5. DT is a *Delaunay triangulation*. It is a graph subdividing the space onto triangles with vertices in $\mathbf{V}$ and edges in $\mathbf{E}$ where the circumcircle of any triangle contains no points of $\mathbf{V}$ other then its vertices.

Each string in the Toussaint hierarchy could be interpreted as a syllogistic proposition. At the beginning, we have just data points without strings (fig.2(a)). Then some first strings have grown up (fig.2(b)) and in some cases we see the first syllogistic conclusions, when three or more points are connected by protoplasmic tubes. At the end, we observe all possible syllogistic conclusions (fig.2(f)) in the topology of attractants and repellents we have set up for *Physarum*. As a result, the growth of plasmodia is considered syllogistic conclusions. Due to different stimuli, we can manage this growth in directions we want, therefore we can foresee all possible syllogistic conclusions, which can be implemented within the certain topology of attractants and repellents. Moreover, we can deal with different syllogistic systems in managing the *Physarum* behaviour. In particular, we can implement the Aristotelian syllogistic (section 3), performative syllogistic (section 4), and Talmudic reasoning by *qal wa-homer* (section 5). For denoting all possible logic circuits of syllogistic systems implemented in the *Physarum* behaviour, we will use the so-called *Physarum diagrams* which are a modification of the well-known Lewis Carroll's diagrams [3], [4].

4 *Physarum* Aristotelian Syllogistic

4.1 Plasmodium without repellents

The Aristotelian syllogistic is the first formal system, it was created in Ancient time. Its axiomatization was laid first by Łukasiewicz [8]. In his axiomatization, the alphabet consists of the syllogistic letters $S, P, M, \ldots$, the syllogistic connectives a, e, i, o, and the propositional connectives $\neg, \vee, \wedge, \Rightarrow$. Atomic propositions are defined as follows: SxP, where $x \in \{a, e, i, o\}$. All other propositions are defined in the following way: (i) each atomic proposition is a proposition, (ii) if X, Y are propositions, then $\neg X, \neg Y, X \star Y$, where $\star \in \{\vee, \wedge, \Rightarrow\}$, are propositions, also. The axioms proposed by Łukasiewicz are as follows:

$$SaP := (\exists A (A \text{ is } S) \wedge \forall A (A \text{ is } S \Rightarrow A \text{ is } P)); \tag{1}$$

$$SiP := \exists A (A \text{ is } S \wedge A \text{ is } P); \tag{2}$$

$$SeP := \neg(SiP); \tag{3}$$

$$SoP := \neg(SaP); \tag{4}$$

$$SaS; \tag{5}$$

$$SiS; \tag{6}$$

$$(MaP \wedge SaM) \Rightarrow SaP; \qquad (7)$$

$$(MaP \wedge MiS) \Rightarrow SiP. \qquad (8)$$

In the *Physarum* implementation of Aristotelian syllogistic, all data points are denoted by appropriate syllogistic letters as attractants. A data point S is considered empty if and only if an appropriate attractant denoted by S is not occupied by plasmodium. We have syllogistic strings of the form SP with the following interpretation: 'S is P,' and with the following meaning: SP is true if and only if S and P are neighbors and both S and P are not empty, otherwise SP is false. By this definition of syllogistic strings, we can define atomic syllogistic propositions as follows:

'All S are P' (SaP): *In the formal syllogistic*: there exists A such that A is S and for any A, if A is S, then A is P. *In the Physarum model*: there is a plasmodium A and for any A, if A is located at S, then A is located at P.

'Some S are P' (SiP): *In the formal syllogistic*: there exists A such that both 'A is S' is true and 'A is P' is true. *In the Physarum model*: there exists a plasmodium A such that A is located at S and A is located at P.

'No S are P' (SeP): *In the formal syllogistic*: for all A, 'A is S' is false or 'A is P' is false. *In the Physarum model*: for all plasmodia A, A is not located at S or A is not located at P.

'Some S are not P' (SoP): *In the formal syllogistic*: for any A, 'A is S' is false or there exists A such that 'A is S' is true and 'A is P' is false. *In the Physarum model*: for any plasmodia A, A is not located at S or there exists A such that A is located at S and A is not located at P.

Formally, this semantics is defined as follows. Let M be a set of attractants. Take a subset $\|X\| \subseteq M$ of attractants occupied by the plasmodium as a meaning for each syllogistic variable X. Next, define an ordering relation $\subseteq$ on subsets $\|S\|, \|P\| \subseteq M$ as: $\|S\| \subseteq \|P\|$ iff all attractants from $\|P\|$ are reachable for the plasmodium located at the attractants from $\|S\|$. Hence, $\|S\| \cap \|P\| \neq \emptyset$ means that some attractants from $\|P\|$ are reachable for the plasmodium located at the attractants from $\|S\|$ and $\|S\| \cap \|P\| = \emptyset$ means that no attractants from $\|P\|$ are reachable for the plasmodium located at the attractants from $\|S\|$. This gives rise to models $\mathcal{M} = \langle M, \|\cdot\| \rangle$ such that

- $\mathcal{M} \models SaP$ iff $\|S\| \subseteq \|P\|$;

- $\mathcal{M} \models SiP$ iff $\|S\| \cap \|P\| \neq \emptyset$;

- $\mathcal{M} \models SeP$ iff $\|S\| \cap \|P\| = \emptyset$;

- $\mathcal{M} \models p \wedge q$ iff $\mathcal{M} \models p$ and $\mathcal{M} \models q$;

- $\mathcal{M} \models p \vee q$ iff $\mathcal{M} \models p$ or $\mathcal{M} \models q$;

- $\mathcal{M} \models \neg p$ iff it is false that $\mathcal{M} \models p$.

Proposition 1. *The Aristotelian syllogistic is sound and complete relatively to $\mathcal{M}$ if we understand $\subseteq$ as an inclusion relation (it is a well-known result [19]).*

However, relatively to all possible plasmodium behaviours the Aristotelian syllogistic is not complete. Indeed, the relation $\subseteq$ can have the following verification on *Physarum* according to our definitions: $\|S\| \subseteq \|P\|$ and $\|S\| \subseteq \|P'\|$, where $\|P\| \cap \|P'\| = \emptyset$, i.e. all attractants from $\|P\|$ are reachable for the plasmodium located at the attractants from $\|S\|$ and all attractants from $\|P'\|$ are reachable for the plasmodium located at the attractants from $\|S\|$, but between $\|P\|$ and $\|P'\|$ there are no paths. In this case $\subseteq$ is not an inclusion relation and proposition 1 does not hold. Hence, we need repellents to make $\subseteq$ the inclusion relations in all cases.

4.2 Plasmodium with repellents

In the *Physarum* diagrams for verifying all the basic syllogistic propositions, we will use the following four cells: x, y, x', y', where x' means all cells which differ from x, but they are neighbors for y, and y' means all cells which differ from y and are neighbors for x. These cells express appropriate meanings of syllogistic letters. The corresponding universe of discourse will be denoted by means of the following diagram: $\begin{array}{|c|c|} \hline x & y' \\ \hline y & x' \\ \hline \end{array}$. Assume that a black counter denotes an attractant and if it is placed within a cell x, this means that "this Voronoi cell contains an attractant N_x activated and occupied by the plasmodium." It is a verification of the syllogistic letter S_x at cell x. A grey counter denotes a repellent and if it is placed within a cell x, this means that "this Voronoi cell contains a repellent R_x activated and there is no plasmodium in it." It is a verification of a new syllogistic letter $[S_x]$. For the sake of convenience, we will denote S_x by x and $[S_x]$ by $[x]$. Using these counters, we can verify all the basic existence syllogistic propositions in a way analogous, though different to Carroll's diagrams (see fig.4).

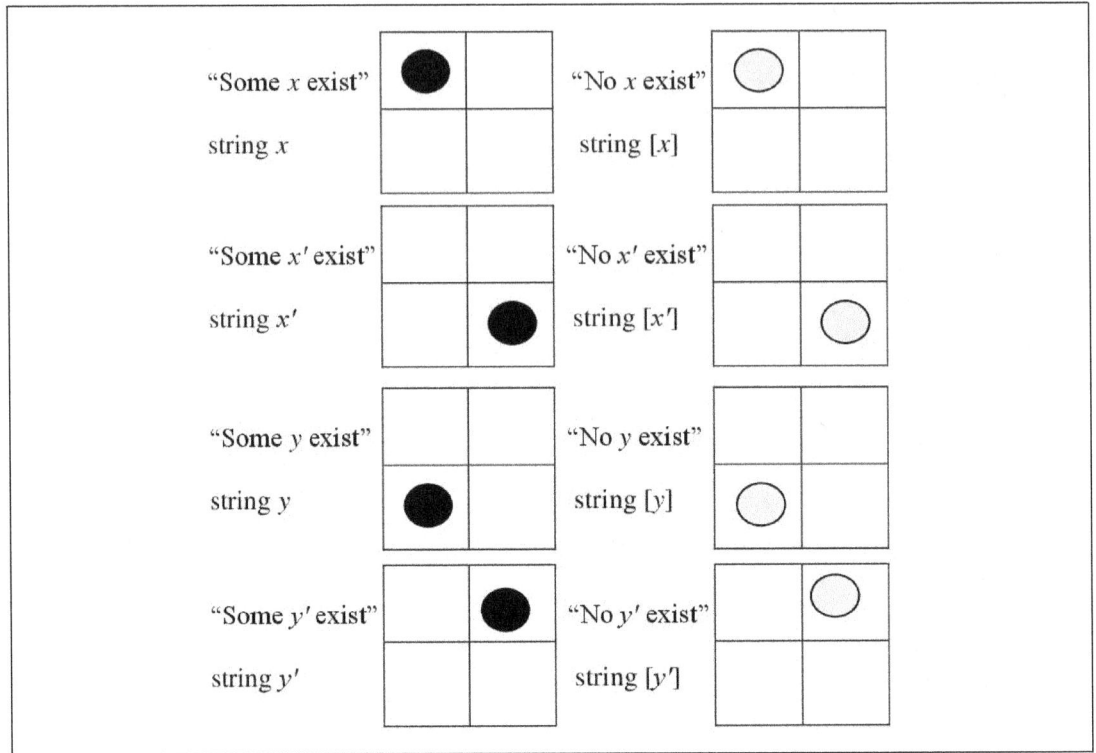

Figure 4: The *Physarum* diagrams for the basic existence strings.

Physarum strings of the form xy, yx are interpreted as particular affirmative propositions "Some x are y" and "Some y are x" respectively, strings of the form $[xy]$, $[yx]$, $x[y]$, $y[x]$ are interpreted as universal negative propositions "No x are y" and "No y are x." A universal affirmative proposition "All x are y" are presented by a complex string $xy \& x[y']$. The sign & means that we have strings xy and $x[y']$ simultaneously and they are considered the one complex string. All these strings are verified on the basis of the diagrams of fig.5.

For verifying syllogisms we will use the following diagrams symbolizing some neighbor cells:

	m	m'	
m'	x	y'	m
m	y	x'	m'
	m'	m	

The motion of plasmodium starts from one of the central cells (x, y, x', y') and goes towards one of the four directions (northwest, southwest, northeast, southeast).

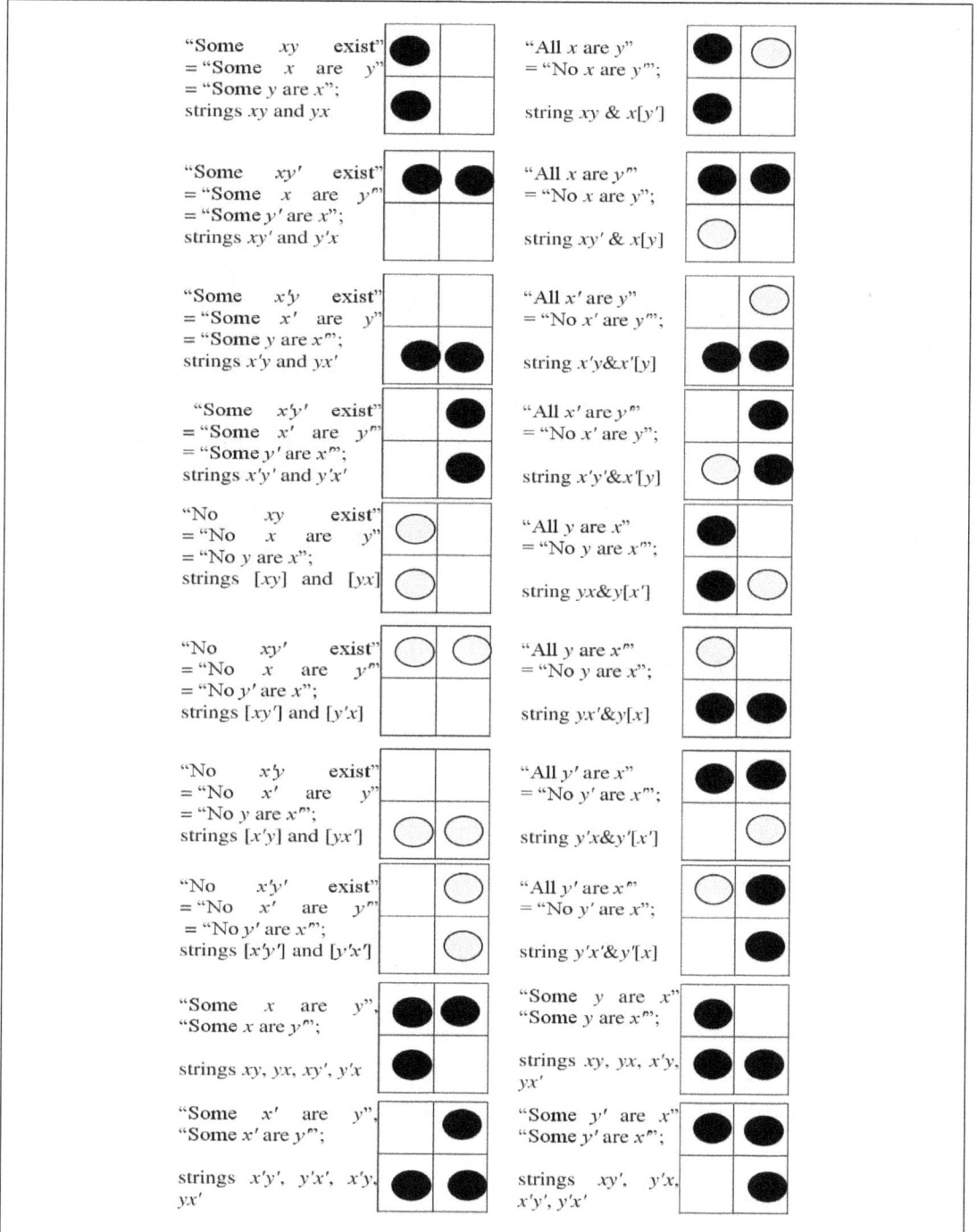

Figure 5: The *Physarum* diagrams for syllogistic propositions.

The syllogism shows a connection between two not-neighbor cells on the basis of its joint neighbor and says if there was either multiplication or fusion. As a syllogistic conclusion, we obtain another diagram:
x	m'
m	x'

Different syllogistic conclusions derived show directions of plasmodium's propagation. Some examples are provided in fig.6–8:

Continuing in the same way, we can construct a syllogistic system, where conclusions are derived from three premises. The motion of plasmodium starts from one of the central cells (x, y, x', y') and goes towards one of the four directions (northwest, southwest, northeast, southeast), then towards one of the eight directions (north-northwest, west-northwest, south-southwest, west-southwest, north-northeast, east-northeast, south-southeast, east-southeast), etc.

Hence, a spatial expansion of plasmodium is interpreted as a set of syllogistic propositions. The universal affirmative proposition $xy\&x[y']$ means that the plasmodium at the place x goes only to y and all other directions are excluded. The universal negative proposition $x[y]$ or $[xy]$ means that the plasmodium at the place x cannot go to y and we know nothing about other directions. The particular affirmative proposition xy means that the plasmodium at the place x goes to y and we know nothing about other directions. Syllogistic conclusions allow us to mentally reduce the number of syllogistic propositions showing plasmodium's propagation.

For the implementation of Aristotelian syllogistic we appeal to repellents to delete some possibilities in the plasmodium propagation. So, model $\mathcal{M}$ defined above should be understood as follows:

- $\mathcal{M} \models$ *All x are y* iff $xy\&x[y']$, i.e. the plasmodium is located at x and can move only to y and cannot move towards all other directions;

- $\mathcal{M} \models$ *Some x are y* iff xy, i.e. the plasmodium is located at x and can move to y;

- $\mathcal{M} \models$ *No x are y* iff $x[y]$ or $[xy]$, i.e. the plasmodium cannot move to y in any case.

It is evident in this formulation that the Aristotelian syllogistic is so unnatural for plasmodia. Without repellents, this syllogistic system cannot be verified in the medium of plasmodium propagations. In other words, we can prove the next proposition:

Proposition 2. *The Aristotelian syllogistic is not sound and complete on the plasmodium without repellents.*

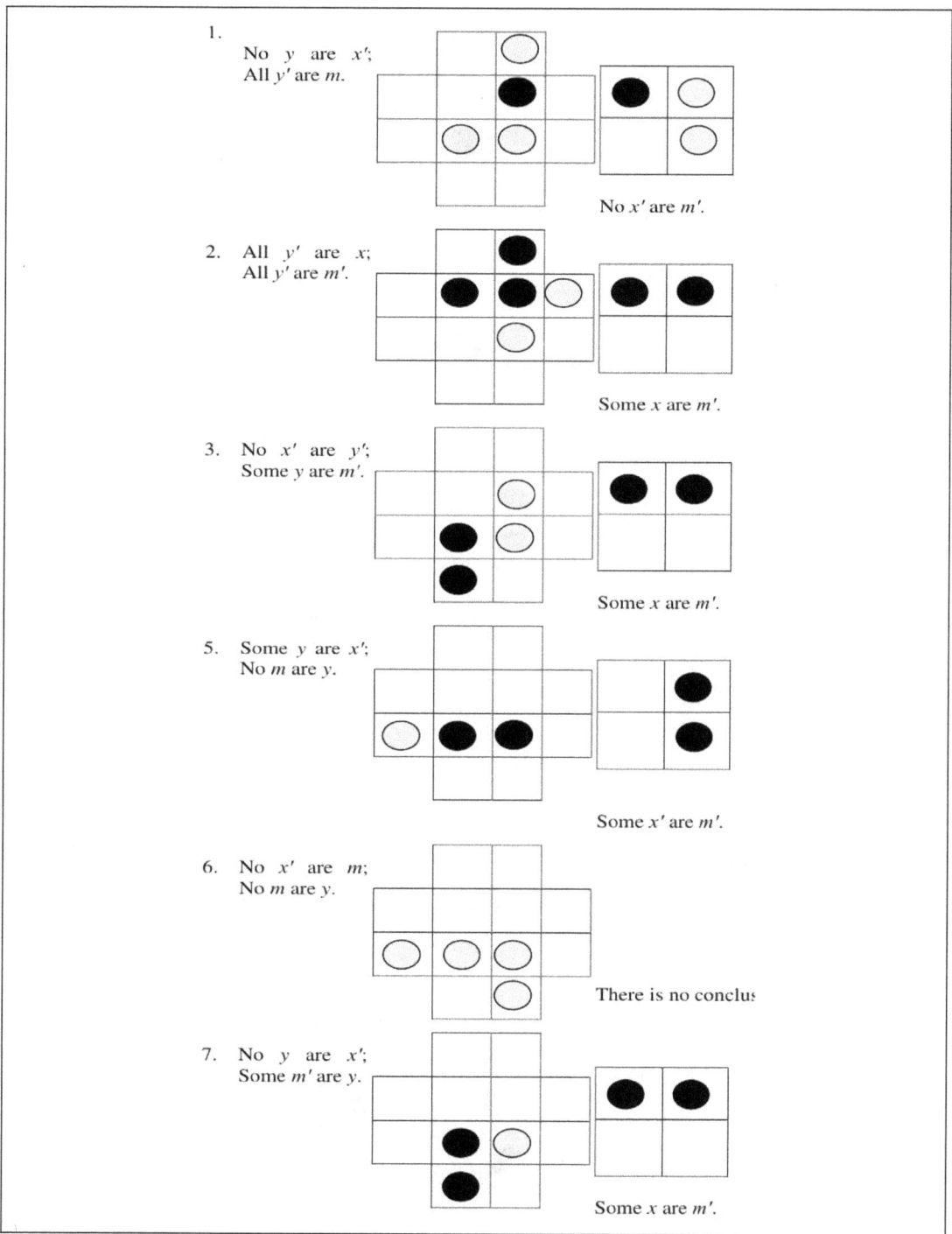

Figure 6: *Physarum* diagrams for syllogisms (part 1).

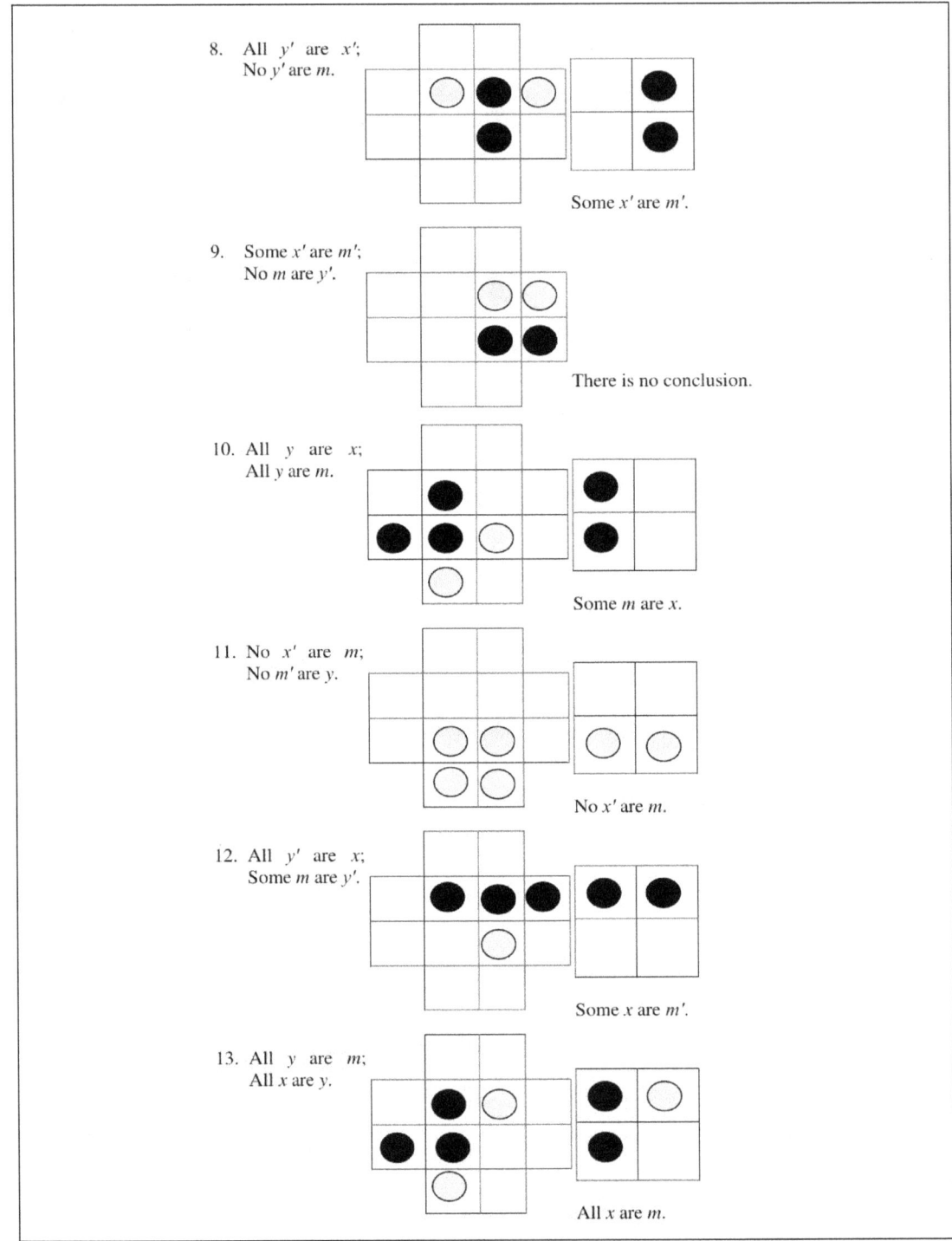

Figure 7: *Physarum* diagrams for syllogisms (part 2).

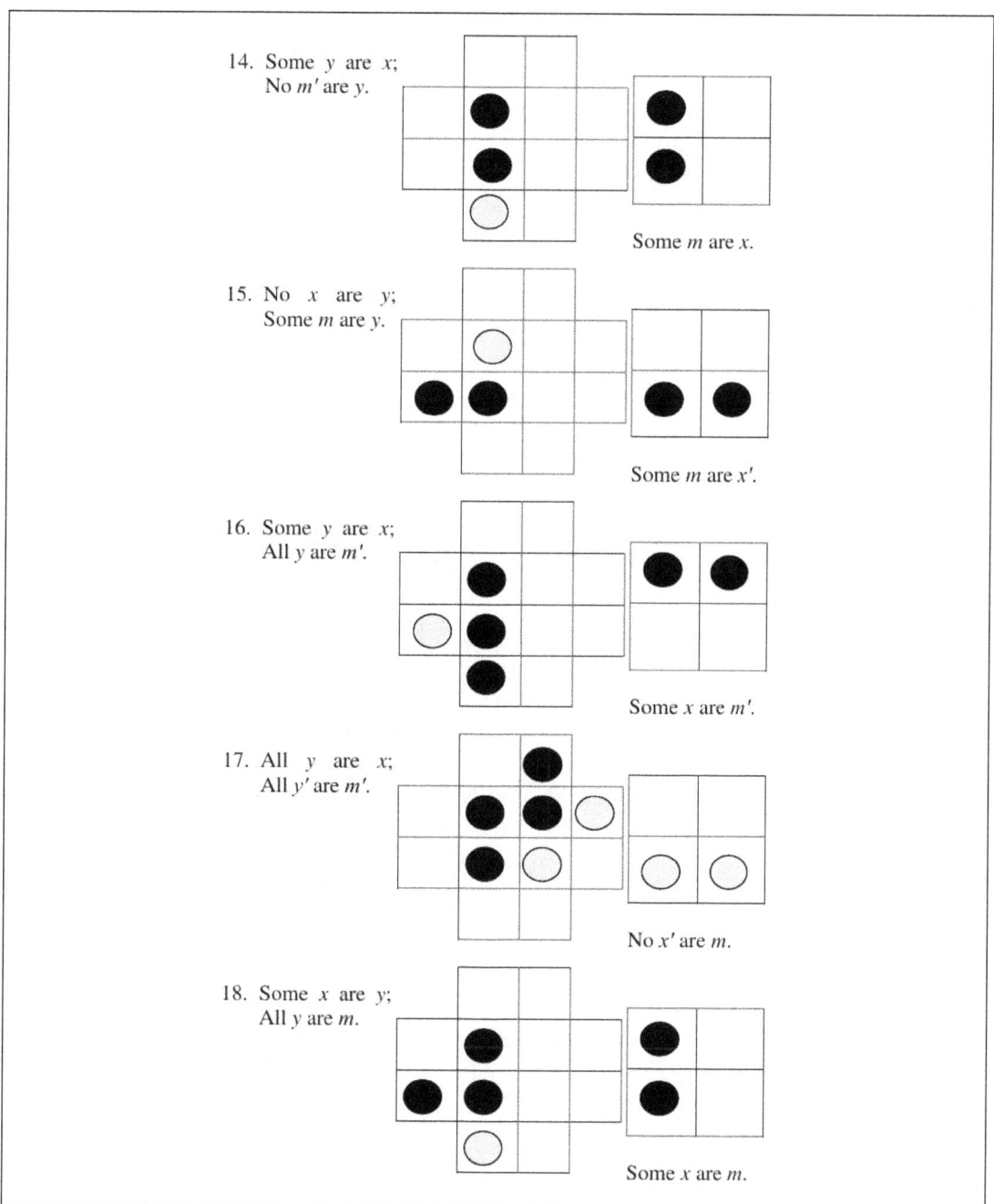

Figure 8: *Physarum* diagrams for syllogisms (part 3).

5   *Physarum* Non-Aristotelian Syllogistic

While in Aristotelian syllogisms we are concentrating on one direction of many *Physarum* motions, and dealing with acyclic directed graphs with fusions of many protoplasmic tubes toward one data point, in most cases of *Physarum* behaviour, not limited by repellents, we observe a spatial expansion of *Physarum* protoplasm in all directions with many cycles, see fig.9. Under these circumstances it is more natural to define all the basic syllogistic propositions SaP, SiP, SeP, SoP in a way they satisfies the inverse relationship when all converses are valid: $SaP \Rightarrow PaS$, $SiP \Rightarrow PiS$, $SeP \Rightarrow PeS$, $SoP \Rightarrow PoS$. In other words, we can draw more natural conclusions for protoplasmic tubes which are decentralized and have some cycles. The formal syllogistic system over propositions with such properties is constructed in [15]. This system is called the *performative syllogistic*. The alphabet of this system contains as descriptive signs the syllogistic letters S, P, M, ..., as logical-semantic signs the syllogistic connectives a, e, i, o and the propositional connectives $\neg$, $\vee$, $\wedge$, $\Rightarrow$. Atomic propositions are defined as follows: SxP, where $x \in \{a, e, i, o\}$. All other propositions are defined thus: (i) each atomic proposition is a proposition, (ii) if X, Y are propositions, then $\neg X$, $\neg Y$, $X \star Y$, where $\star \in \{\vee, \wedge, \Rightarrow\}$, are propositions, too.

In order to implement the performative syllogistic in the behaviour of *Physarum* plasmodium, we will interpret all data points denoted by appropriate syllogistic letters as attractants. A data point S is considered empty if and only if an appropriate attractant denoted by S is not occupied by plasmodium. Let us define syllogistic strings of the form SP with the following interpretation: 'S is P,' and with the following meaning: SP is true if and only if S and P are reachable for each other by the plasmodium and both S and P are not empty, otherwise SP is false. Using this definition of syllogistic strings, we can define atomic syllogistic propositions as follows:

'All *S* are *P*' (SaP): *In the formal performative syllogistic*: there exists A such that A is S and for any A, A is S and A is P. *In the Physarum model*: there is a string AS and for any A which is a neighbor for S and P, there are strings AS and AP. This means that we have a massive-parallel occupation of the region where the cells S and P are located.

'Some *S* are *P*' (SiP): *In the formal performative syllogistic*: for any A, both 'A is S' is false and 'A is P' is false. *In the Physarum model*: for any A which is a neighbor for S and P, there are no strings AS and AP. This means that the plasmodium cannot reach S from P or P from S immediately.

Figure 9: Development of protoplasmic network by the plasmodium. Snapshots (a)–(d) are recorded with 10 hour intervals. In this experiment we observe cycles and decentralization of *Physarum* motions.

'No S are P' (SeP): *In the formal performative syllogistic*: there exists A such that if 'A is S' is false, then 'A is P' is true. *In the Physarum model*: there exists A which is a neighbor for S and P such that there is a string AS or there is a string AP. This means that the plasmodium occupies S or P, but not the whole region where the cells S and P are located.

'Some S are not P' (SoP): *In the formal performative syllogistic*: for any A, 'A

is S' is false or there exists A such that 'A is S' is false or 'A is P' is false. *In the Physarum model*: for any A which is a neighbor for S and P there is no string AS or there exists A which is a neighbor for S and P such that there is no string AS or there is no string AP. This means that the plasmodium does not occupy S or there is a neighboring cell which is not connected to S or P by a protoplasmic tube.

Composite propositions are defined in the standard way.
In the performative syllogistic we have the following axioms:

$$SaP := (\exists A(A \text{ is } S) \land (\forall A(A \text{ is } S \land A \text{ is } P))); \tag{9}$$

$$SiP := \forall A(\neg(A \text{ is } S) \land \neg(A \text{ is } P)); \tag{10}$$

$$SoP := \neg(\exists A(A \text{ is } S) \lor (\forall A(A \text{ is } P \land A \text{ is } S))), i.e.$$
$$(\forall A \neg(A \text{ is } S) \land \exists A(\neg(A \text{ is } P) \lor \neg(A \text{ is } S))); \tag{11}$$

$$SeP := \neg \forall A(\neg(A \text{ is } S) \land \neg(A \text{ is } P)), i.e.$$
$$\exists A(A \text{ is } S \lor A \text{ is } P). \tag{12}$$

$$SaP \Rightarrow SeP; \tag{13}$$

$$SaP \Rightarrow PaS; \tag{14}$$

$$SiP \Rightarrow PiS; \tag{15}$$

$$SaM \Rightarrow SeP; \tag{16}$$

$$MaP \Rightarrow SeP; \tag{17}$$

$$(MaP \land SaM) \Rightarrow SaP; \tag{18}$$

$$(MiP \land SiM) \Rightarrow SiP. \tag{19}$$

The formal properties of this axiomatic system are considered in [15]. In the performative syllogistic we can analyze the collective dimension of behaviour. Within this system we can study how the plasmodium occupies all possible attractants in any direction if it can see them. So, this system shows logical properties of a massive-parallel behaviour (i.e. the collective dimension of behaviour). One of the most significant notions involved in this implementation of the performative syllogistic in *Physarum* topology is a *neighborhood*. We can define a distance for the neighborhood differently, i.e. we can make it broader or narrower. So, from different neighborhoods it will follow that we deal with different 'universes of discourse.'

In the *Physarum* diagrams for the performative syllogistic, the 'universe of discourse' cover cells x, y, non-x (which be denoted by x'), non-y (which he denoted by y'):

x	y'
y	x'

, where x, y, x', y' are neighbor cells containing attractants for *Physarum*, x' are all neighbors for y which differ from x, and y' are all neighbors for x which differ from y. Suppose that we have black, white, and grey counters and (i) if a black counter is placed within a cell, this means that "this cell is occupied" (i.e. "there is at least one thing in it"), (ii) if a white counter is placed within a cell, this means that "this cell is not occupied" (i.e. "there is not thing in it"), (ii) if a grey counter is placed within a cell, this means that "it is not known if this cell is occupied". All possible combinations of *Physarum* diagrams for atomic propositions within our universe of discourse are pictured in fig.10.

The universe of discourse for simulating performative syllogisms by means of *Physarum* behaviours covers cells x, y, m, x', y', m' in the following manner:

y'	m	m'	x'
m'	x	y'	m
m	y	x'	m'
x	m'	m	y

The motion of plasmodium starts from one of the central cells (x, y, x', y') and goes towards one of the four directions (northwest, southwest, northeast, southeast). The *Physarum* diagram for syllogistic conclusions is as follows:

x	m'
m	x'

Some examples of performative syllogistic conclusions are regarded in fig.11. A zone of true universal affirmative propositions is pictured in fig.12.

Thus, the performative syllogistic allows us to study different zones containing attractants for *Physarum* if they are connected by protoplasmic tubes homogenously as in fig.12.

A model $\mathcal{M}' = \langle M', \|\cdot\|_x \rangle$ for the performative syllogistic, where M' is the set of attractants and $\|X\|_x \subseteq M'$ is a meaning of syllogistic letter X which is understood as all attractants reachable for the plasmodium from the point x, is defined as follows:

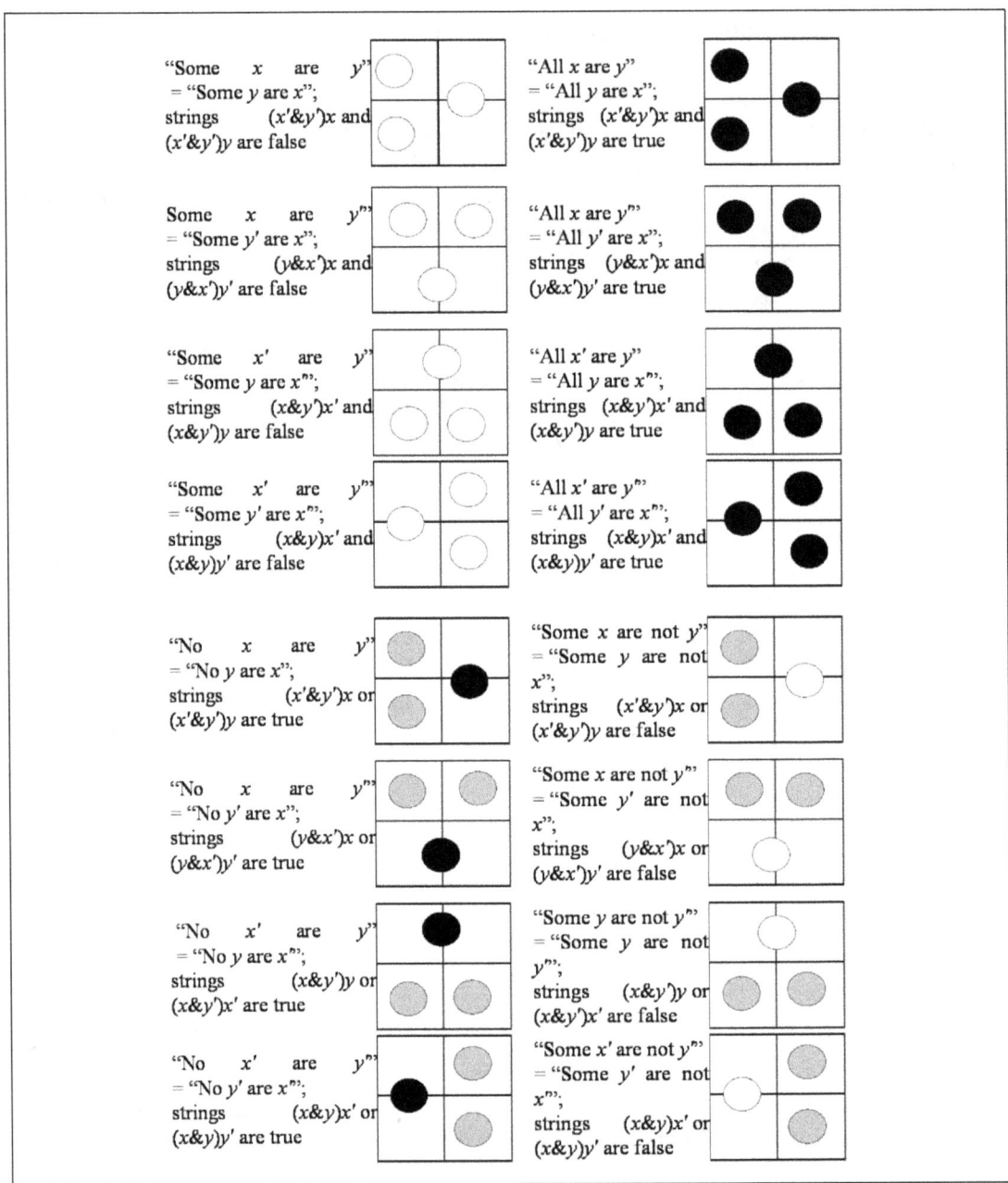

Figure 10: The *Physarum* diagrams for premises of performative syllogisms. Strings of the form $(x'\&y')x$ mean that in cells x' and y' there are neighbors A for x such that Ax, i.e. $(x'\&y')$ is a metavariable in $(x'\&y')x$ that is used to denote all attractants of x' and y' which are neighbors for the attractant of x.

DIAGRAMS FOR SYLLOGISMS

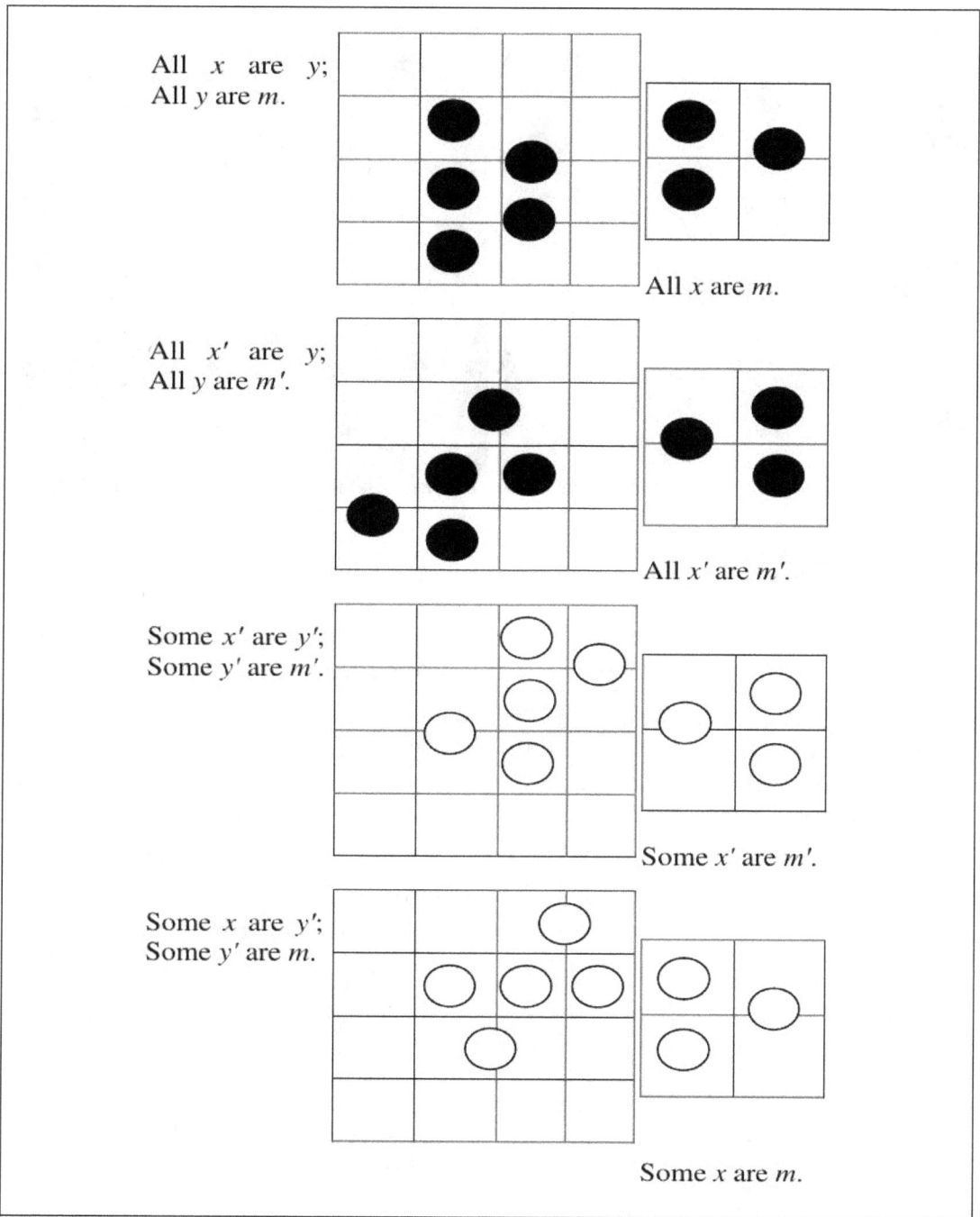

Figure 11: The *Physarum* diagrams for performative syllogisms with true conclusions.

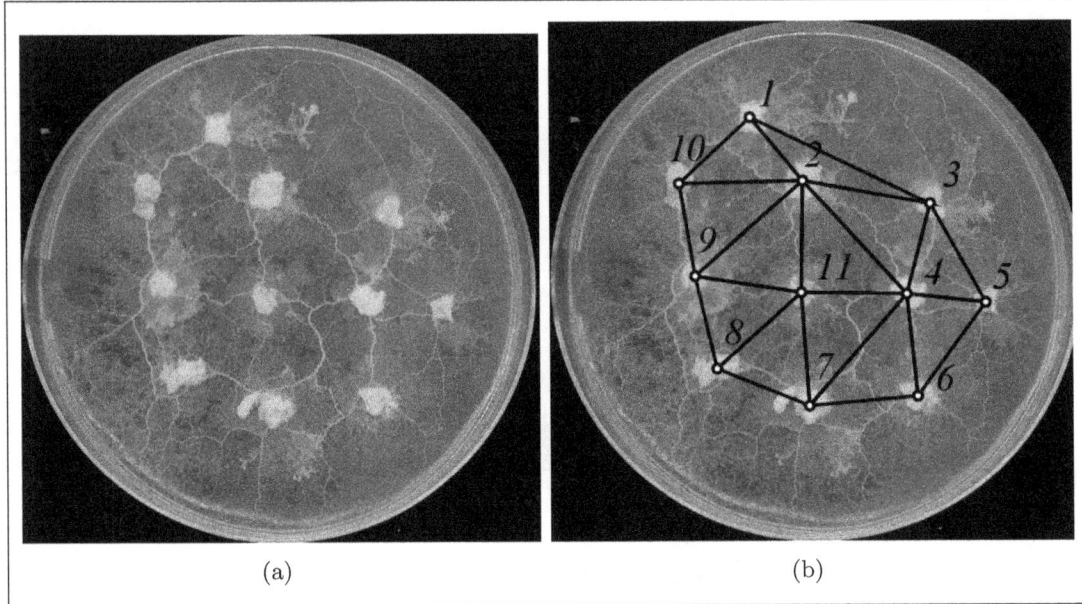

Figure 12: The active zone, where all cells containing attractants are connected by protoplasmic tubes, therefore the following syllogistic propositions are valid: $S_1 a S_{10}$, $S_2 a S_3$, $S_9 a S_{11}$, $S_7 a S_6$, etc. and the following syllogistic conclusions are true: $((S_{11} a S_9 \wedge S_9 a S_{10}) \Rightarrow S_{11} a S_{10})$, $((S_1 a S_2 \wedge S_2 a S_4) \Rightarrow S_1 a S_4)$, etc.

- $\mathcal{M}' \models$ All x are y iff $\|X\|_x \neq \emptyset$, $\|X\|_y \neq \emptyset$, and $\|X\|_x \cap \|X\|_y \neq \emptyset$, more precisely both $(x'\&y')x$ and $(x'\&y')y$ hold in $\mathcal{M}'$, i.e. the plasmodium can move from neighbors of y to x and it can move from neighbors of x to y;

- $\mathcal{M}' \models$ Some x are y iff $y \notin \|X\|_x$ and $x \notin \|X\|_y$, more precisely neither $(x'\&y')x$ nor $(x'\&y')y$ hold in $\mathcal{M}'$, i.e. the plasmodium cannot move from neighbors of y to x and it cannot move from neighbors of x to y;

- $\mathcal{M}' \models$ No x are y iff $y \in \|X\|_x$ or $x \in \|X\|_y$, more precisely $(x'\&y')x$ or $(x'\&y')y$ hold in $\mathcal{M}'$, i.e. the plasmodium can move from neighbors of y to x or it can move from neighbors of x to y;

- $\mathcal{M}' \models$ Some x are not y iff $y \notin \|X\|_x$ or $x \notin \|X\|_y$, more precisely $(x'\&y')x$ or $(x'\&y')y$ do not hold in $\mathcal{M}'$, i.e. the plasmodium cannot move from neighbors of y to x or it cannot move from neighbors of x to y;

- $\mathcal{M}' \models p \wedge q$ iff $\mathcal{M}' \models p$ and $\mathcal{M}' \models q$;

- $\mathcal{M}' \models p \vee q$ iff $\mathcal{M}' \models p$ or $\mathcal{M}' \models q$;

- $\mathcal{M}' \models \neg p$ iff it is false that $\mathcal{M}' \models p$.

Proposition 3. *The performative syllogistic is sound and complete in $\mathcal{M}'$.*

For more details on formal properties of performative syllogistic, please see [15]. This syllogistic describes the logic of plasmodium propagation in all possible directions. For the implementation of this syllogistic we do not need repellents. It is a natural system.

6 *Physarum* Talmudic Reasoning

Deductions in Talmudic reasoning are constructed by using different inference rules. The oldest family of these rules consists of the thirteen rules of Rabbi Ishmael. The most important rule among them is called *qal wa-homer*. The direct meaning of the word *qal* is 'light in weight.' From a logical point of view, *qal* is regarded as *minor*, i.e. as being less important or less significant. The direct meaning of *homer* is 'heaviness.' It is major, i.e. more important, more significant. Hence, *qal wa-homer* is an inference from minor to major, and vice versa, from major to minor. For example, the Sabbath day is in some respects regarded as minor of a common holiday or festival (*yom tov*). Therefore, if a certain kind of work is permitted on the Sabbath we can infer that such a work is the more permissible on a festival; and vice versa, if a certain work is forbidden on festival, it has to be forbidden on the Sabbath also. Thus, *qal wa-homer* concerns actions when they are permitted to be performed.

Notice that in the Talmud it is claimed that it is sufficient to derive from an inference a result that is equivalent to the law from which it is made, i.e. the law transferred to the major must never surpass in severity the original law in the minor, from which the inference was drawn. This way of thinking is said to be the *dayo* principle.

Let us consider a Talmudic example of *qal wa-homer*. In the *Baba Qama* (one of the Talmudic books), different kinds of damages (*nezeqin*) are analyzed, among which three genera are examined: foot action (*regel*), tooth action (*shen*) and horn action (*qeren*). These three are damages that could be caused by an ox (he can trample (foot), eat (tooth) and gore (horn)). Due to the Torah it is known that tooth damage (as well as foot damage) by an ox at a public place needs to pay zero compensation. Horn damage at a public place entails payment of 50% the damage costs as compensation. In a private area foot/tooth damage must be paid in full. What can we say now about payments for horn actions of private places?

Damages (nezeqin)	Public place	Private place
Horn action (*qeren*)	50%	?
Foot action (*regel*)	0	100%
Tooth action (*shen*)	0	100%

In order to draw up a conclusion by *qal wa-homer*, we should define a two-dimensional ordering relation on the set of data: (i) on the one hand, according to the *dayo* principle, we know that payment for horn action in a private area cannot be greater than the same in a public area, (ii) on the other hand, payment for horn action at a private place cannot be greater than foot/tooth action at the same place. Hence, we infer that payment of compensation for horn action at a private place is equal to 50% of the damage costs.

The table above shows us that *qal wa-homer* can be interpreted spatially, too. Already Yisrael Ury [23] proposed to use the Carroll's bilateral diagrams for modelling conclusions by *qal wa-homer*. Let us take now the diagram $\begin{array}{|c|c|} \hline xy' & xy \\ \hline x'y' & x'y \\ \hline \end{array}$ that plays the role of 'universe of discourse' for Talmudic reasoning over adjuncts x, y, non-x in the neighborhood (that is denoted by x'), non-y in the neighborhood (that is denoted by y'). Assume that we have only black counters and if a black counter is placed within a cell, this means that "this cell is occupied" (i.e. "there is at least one thing in it", "an appropriate Talmudic rule should be obeyed"). Thus, the cell that does not contain a black counter indicates a situation in which the obligation is not fulfilled, whereas the cell containing a black counter indicates a situation in which the obligation is fulfilled.

Hence, if we have two rows and two columns, there are sixteen possible ways to cover such a diagram by means of black counters, in connect to Yisrael Ury who accepts only six of them (fig.13).

Let x mean proposition 1 and y mean proposition 2. Then the above mentioned accepted diagrams (fig.13) have the following sense: (a) It is necessary and sufficient to obey x; (b) It is necessary and sufficient to obey y; (c) It is sufficient to obey either x or y; (d) It is not sufficient to obey x and/or y; (e) It is necessary and sufficient to obey both x and y; (f) It is not necessary to obey either x or y.

Let us return to our example. Let x_3 be a 'foot action', x_2 a 'tooth action', x_1 a 'horn action', y' 'at public place', y 'at private place'. So, we obtain the following diagram:

$x_1 y'$	$x_1 y$
$x_2 y'$	$x_2 y$
$x_3 y'$	$x_3 y$

Assume that a black counter means an obligation to pay 100% of the damage costs as compensation and a grey counter means an obligation to pay any 50%. We

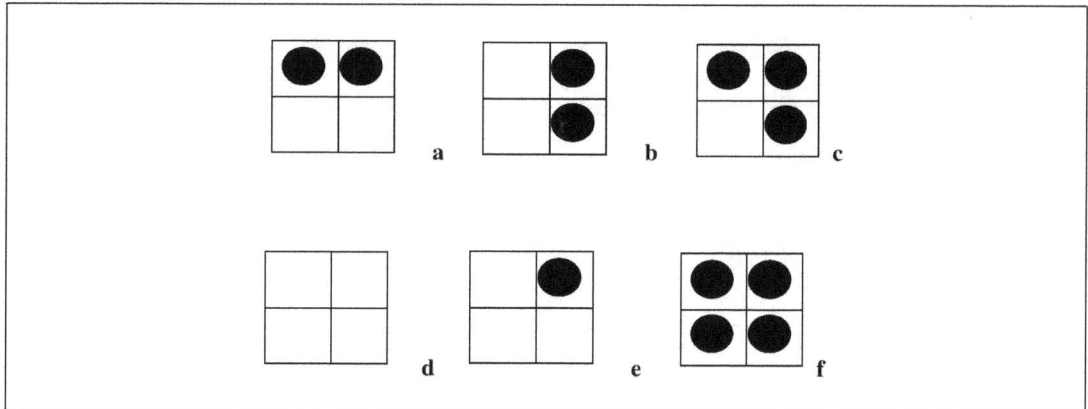

Figure 13: Ury's diagrams for conclusions by *qal wa-homer*.

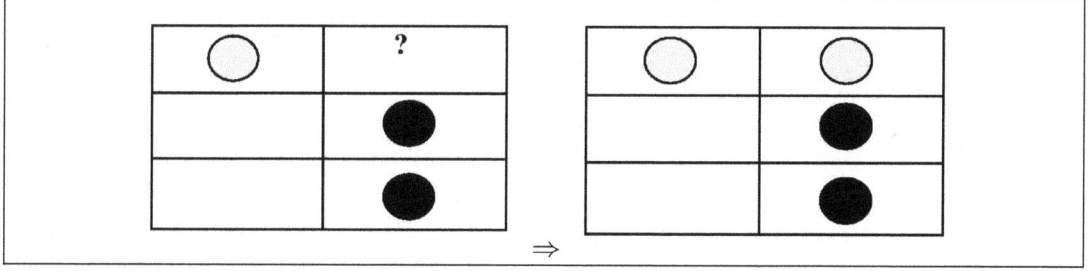

Figure 14: Ury's diagrams for inferring whether we should pay a horn action at a private place.

can cover this diagram by counters as shown in fig.14.

In so doing, we have supposed that there is a different power of intensity in obligation. In this case our rule for inferring by *qal wa-homer* is formulated thus: *if a cell contains a black or grey counter, all cells above it and to its right also contain a black or grey counter and the color of that counter has the minimal hardness of black and grey in counters of the neighbor cells; if a cell does not contain any counter, all cells below it and to its left are also without counters.*

Talmudic diagrams defined above are closer to Carroll's diagrams and are more natural for *Physarum* behaviour than Ury's diagrams (cf. [23]). In the *Physarum* topology, Talmudic diagrams are built on syllogistic strings of the form xy, yx, $x'y$, yx', xy', $y'x$, $x'y'$, $y'x'$, where x and y in xy are interpreted as two neighbor attractants connected by protoplasmic tubes, x' is understood as all attractants which differ from x, but are neighbors of y, and y' is understood as all attractants which differ from y and are neighbors of x. The Talmudic diagrams for the *Physarum*

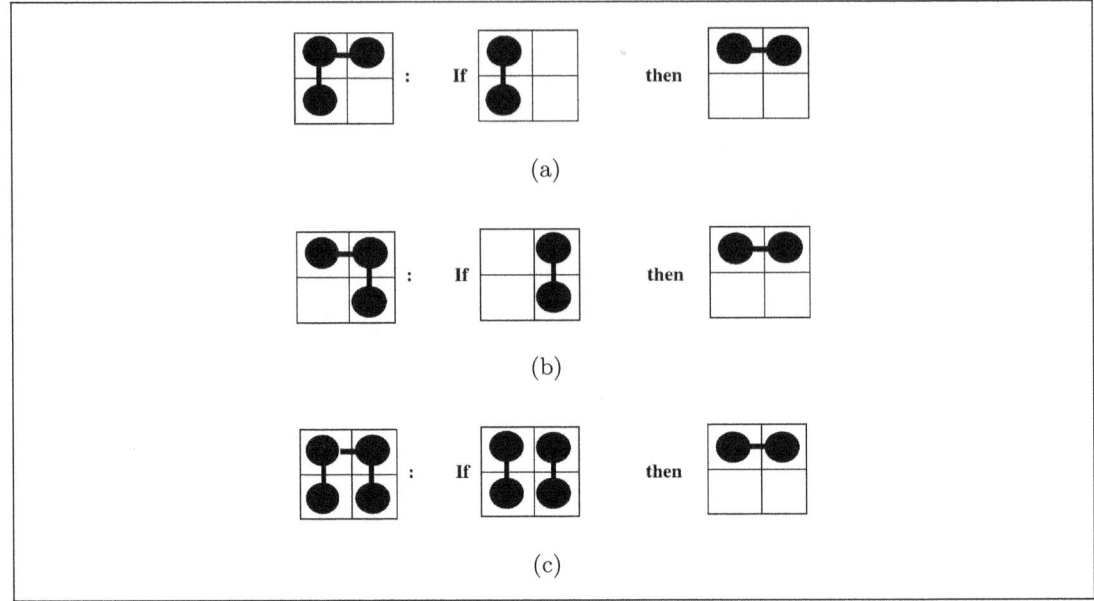

Figure 15: *The Physarum diagrams for qal wa-homer syllogisms* (part 1): (a) If the string xy' is verified, then the string xy is verified, too (i.e. if xy' is verified, then x has a multiplication of plasmodium). (b) If the string yx' is verified, then the string xy is verified, too (i.e. if yx' is verified, then y has a multiplication of plasmodium). (c) If the strings xy' and yx' are verified, then the string xy is verified, too (i.e. if both xy' and yx' are verified, then both x and y have multiplications of plasmodium).

simulation have the following form $\begin{array}{|c|c|} \hline x & y \\ \hline y' & x' \\ \hline \end{array}$ where x' is a non-empty class of neighbor attractants for y and y' is a non-empty class of neighbor attractants for x. Then *qal wa-homer* tells us whether a multiplication took place during the plasmodium's propagation at points x and/or y. In fig.15, all the possible conclusions inferred by *qal wa-homer* in relation to x and y are considered, and they are defined if we have a multiplication at those points.

Hence, the main difference between the Aristotelian syllogistic and Talmudic reasoning is that, on the one hand, we are concentrating on fusions of plasmodium in the case of the Aristotelian syllogistic in the *Physarum* topology and, on the other hand, we deal with multiplications of plasmodia in the case of Talmudic reasoning. An example of an experiment with *Physarum* that satisfies the *qal wa-homer* rule is shown in fig.17. Talmudic reasoning can describe only fragments of

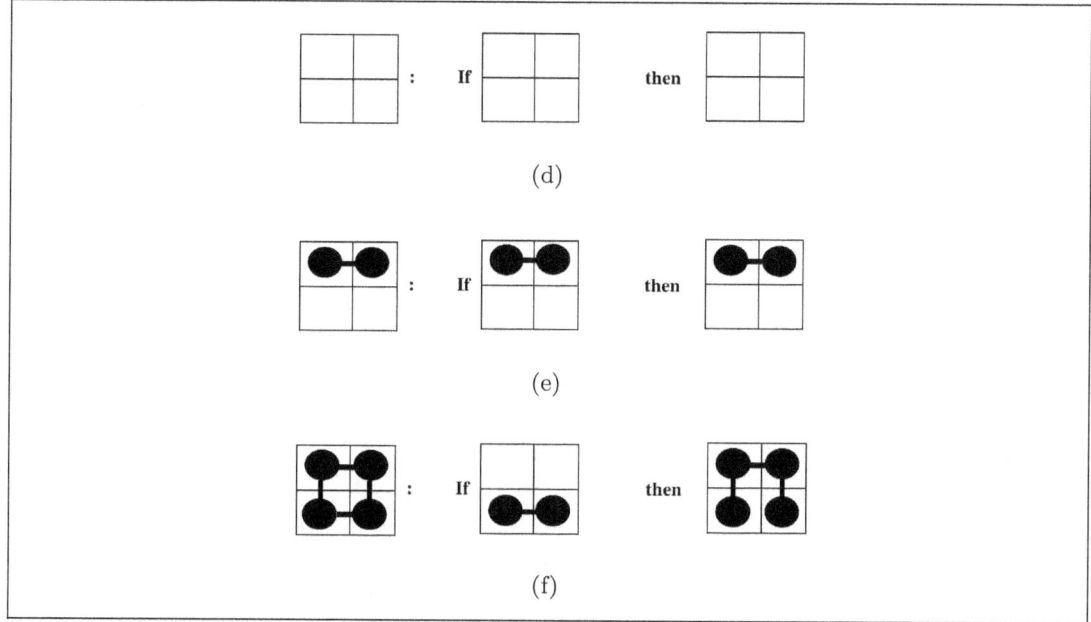

Figure 16: *The Physarum diagrams for qal wa-homer syllogisms* (part 2): (d) If no string is verified, then there is no multiplication of plasmodium. (e) If the string xy is verified, then the there is no multiplication of plasmodium. (f) If the strings $x'y'$ is verified, then the strings xy', $x'y$, xy are verified, too (i.e. if $x'y'$ is verified, then both x and y have multiplications of plasmodium).

plasmodium behaviours while the Aristotelian syllogistic describes some fragments of plasmodium propagations. Only the performative syllogistic is sound and complete on plasmodium interactions. For more details on Talmudic reasoning and its modern formalizations see [17].

7 Conclusion

In this paper, we have considered bio-inspired implementations of several spatial syllogistic calculi. The medium of implementations chosen is the plasmodium of *Physarum Polycephalum*. These implementations are called *Physarum diagrams for syllogistic*. One of the most interesting results is that the Aristotelian syllogistic is quite unnatural in the sense that this system assumes fusions and concentrations of all plasmodium motions in one conclusion. This is difficult for the plasmodium, as it aims to be propagated in all possible directions.

The main theoretical result of our paper is to demonstrate that the performa-

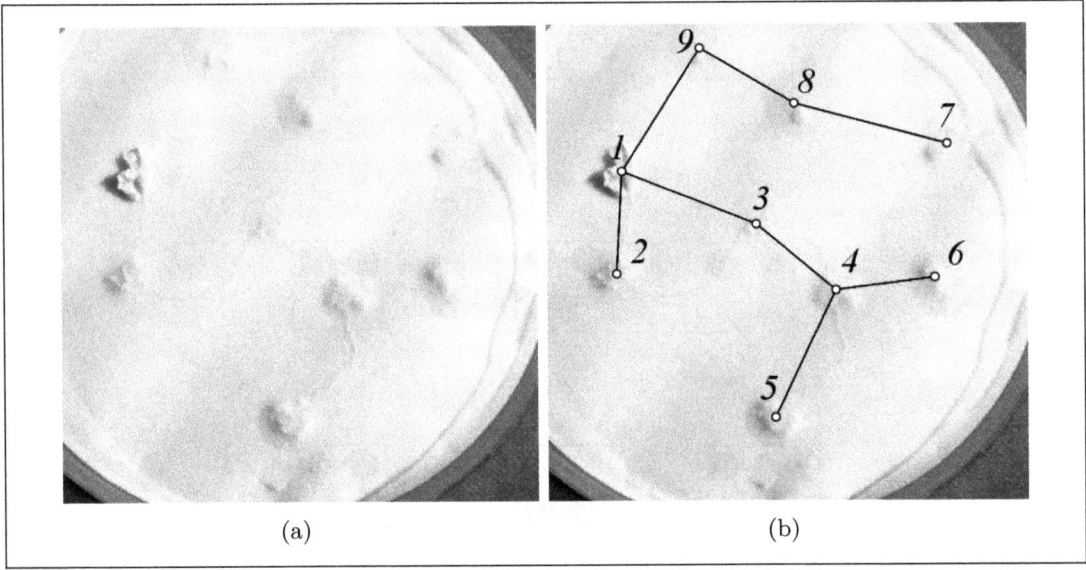

Figure 17: The *qal wa-homer* with plasmodium (part 1): (a) position of oat flakes representing data points, (b) the plasmodium tree corresponding to *qal wa-homer*, where $x = 1$, $y = 9$, $x' = 8$, $y' = 2 \cup 3$, etc.,

tive syllogistic of [15] can simulate massive-parallel behaviours of living organisms such as plasmodia. This result can find many applications in behavioural sciences, because the plasmodium behaviour can be considered the simplest natural intelligent behaviour solving complex tasks. Thus, our result may have an impact on computational models.

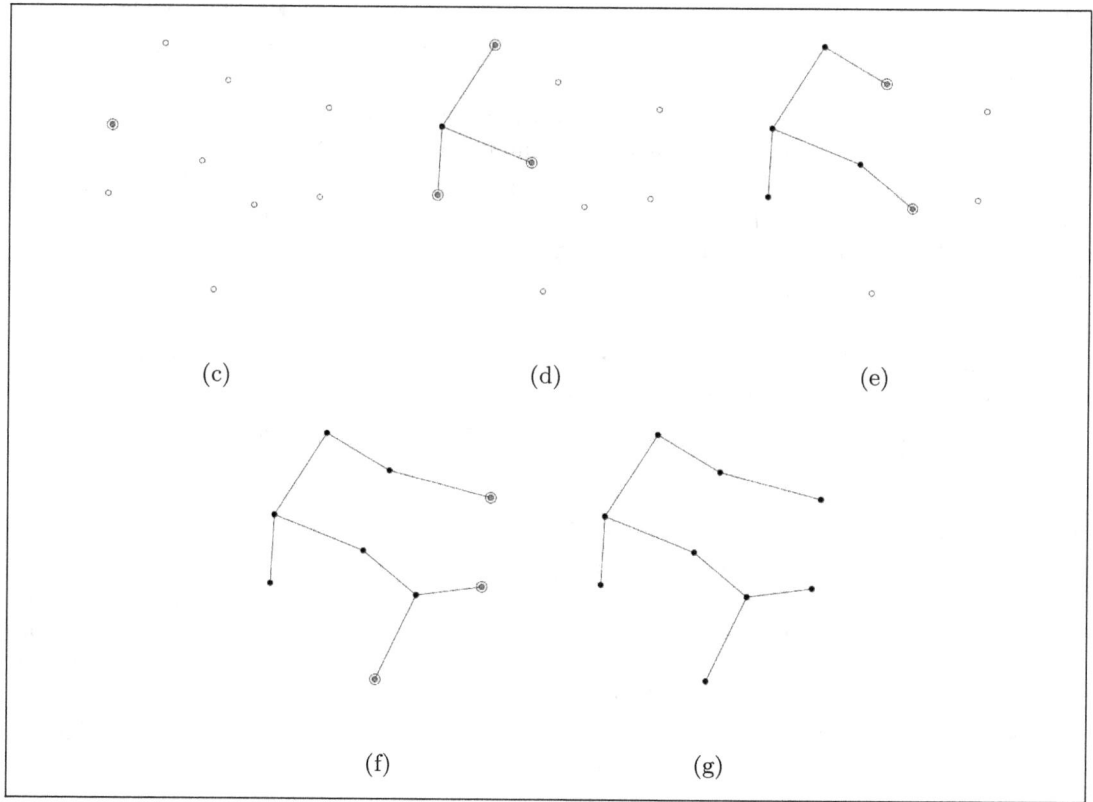

Figure 18: The *qal wa-homer* with plasmodium (part 2): (c)–(g) dynamics of *qal wa-homer* by the growing plasmodium.

References

[1] Adamatzky, A., Physarum machine: implementation of a Kolmogorov-Uspensky machine on a biological substrate, *Parallel Processing Letters*, 17(4):455–467, 2007.

[2] Adamatzky, A., *Physarum Machines: Computers from Slime Mould* (World Scientific Series on Nonlinear Science, Series A). World Scientific Publishing Company, 2010.

[3] Carroll, L. *The Game of Logic*. London: Macmillan and co., 1886.

[4] Carroll, L. *Symbolic Logic. Part I. Elementary*. London: Macmillan and co., 1897.

[5] Chopard, B., and M. Droz, *Cellular Automata Modeling of Physical Systems*. Cambridge University Press, 2005.

[6] Gutiérrez, A., Marco, S. (eds.), *Biologically Inspired Signal Processing for Chemical Sensing*. Series: Studies in Computational Intelligence, Vol. 188, XIII, 2009.

[7] Kolmogorov, A. N. On the concept of algorithm. *Uspekhi Mat Nauk*, 8(4):175–176, 1953.

[8] Łukasiewicz, J. Aristotle's Syllogistic From the Standpoint of Modern Formal Logic. Oxford Clarendon Press, 2nd edition, 1957.

[9] T. Nakagaki, H. Yamada, A. Toth. Maze-solving by an amoeboid organism. *Nature* 407:470–470, 2000.

[10] T. Nakagaki, H. Yamada, and A. Tothm. Path finding by tube morphogenesis in an amoeboid organism. *Biophysical Chemistry*, 92:47–52, 2001.

[11] T. Nakagaki, M. Iima, T. Ueda, Y. Nishiura, T. Saigusa, A. Tero, R. Kobayashi, K. Showalter. Minimum-risk path finding by an adaptive amoeba network. *Physical Review Letters*, 99:68–104, 2007.

[12] T. Saigusa, A. Tero, T. Nakagaki, Y. Kuramoto. Amoebae Anticipate Periodic Events. *Phys. Rev. Lett*, 100(1), 2008.

[13] Schumann, A., Akimova, L., Simulating of Schistosomatidae (Trematoda: Digenea) behaviour by Physarum Spatial Logic, *Annals of Computer Science and Information Systems, Volume 1. Proceedings of the 2013 Federated Conference on Computer Science and Information Systems*. IEEE Xplore, 2013, 225–230.

[14] Schumann, A., Khrennikov, A., p-Adic Physics, Non-well-Founded Reality and Unconventional Computing, *P-Adic Numbers, Ultrametric Analysis, and Applications* 1(4):297–306, 2009.

[15] Schumann, A., On Two Squares of Opposition: the Lesniewski's Style Formalization of Synthetic Propositions, *Acta Analytica* 28:71–93, 2013.

[16] Schumann, A., Pancerz, K., Towards an Object-Oriented Programming Language for Physarum Polycephalum Computing, [in:] M. Szczuka, L. Czaja, M. Kacprzak (eds.), *Proceedings of the Workshop on Concurrency, Specification and Programming* (CS&P'2013), Warsaw, Poland, September 25-27:389–397, 2013.

[17] Schumann, A. (ed.), Modern Review of Judaic Logic. *Special issue of History and Philosophy of Logic*, 2011.

[18] T. Shirakawa, Y.-P. Gunji, and Y. Miyake. An associative learning experiment using the plasmodium of *Physarum polycephalum. Nano Communication Networks*, 2:99–105,

2011.

[19] Smith, R. Completeness of an ecthetic syllogistic. *Notre Dame J. Formal Logic*, 24(2):224–232, 1983.

[20] A. Tero, T. Nakagaki, K. Toyabe, K. Yumiki, R. Kobayashi. A Method Inspired by Physarum for Solving the Steiner Problem. *IJUC*, 6(2):109–123, 2010.

[21] S. Tsuda, M. Aono, and Y.P. Gunji. Robust and emergent Physarum-computing. *BioSystems*, 73:45–55, 2004.

[22] Uspensky, V. U. Kolmogorov and mathematical logic. *J Symb Logic*, 57:385–412, 1992.

[23] Ury, Yisrael. *Charting the Sea of Talmud: A Visual Method for Understanding the Talmud*, 2012.

[24] Sh. Watanabe, A. Tero, A. Takamatsu, T. Nakagaki. Traffic optimization in railroad networks using an algorithm mimicking an amoeba-like organism, Physarum plasmodium. *Biosystems*, 105(3):225–232, 2011.

TABDUAL: A TABLED ABDUCTION SYSTEM FOR LOGIC PROGRAMS

ARI SAPTAWIJAYA*
NOVA Laboratory for Computer Science and Informatics (NOVA LINCS)
Departamento de Informática, Faculdade de Ciências e Tecnologia
Universidade Nova de Lisboa, 2829-516 Caparica, Portugal
`ar.saptawijaya@campus.fct.unl.pt`

LUÍS MONIZ PEREIRA
NOVA Laboratory for Computer Science and Informatics (NOVA LINCS)
Departamento de Informática, Faculdade de Ciências e Tecnologia
Universidade Nova de Lisboa, 2829-516 Caparica, Portugal
`lmp@fct.unl.pt`

Abstract

Abduction has been on the back burner in logic programming, as abduction can be too difficult to implement, and costly to perform, in particular if abductive solutions are not tabled. On the other hand, current Prolog systems, with their tabling mechanisms, are mature enough to facilitate the introduction of tabling abductive solutions (tabled abduction) into them.

Our contributions are as follows. First, we conceptualize a tabled abduction technique for abductive normal logic programs, permitting abductive solutions to be reused, from one abductive context to another. The approach is underpinned by the theory of ABDUAL and relies on a transformation into tabled logic programs. It particularly makes use of the dual transformation of ABDUAL that enables efficiently handling the problem of abduction under negative goals, by introducing dual positive counterparts for them. Second, we realize this tabled abduction technique in TABDUAL, a system implemented in XSB Prolog. The implementation poses several challenges to concretely realize the abstract theory of ABDUAL, e.g., by taking care of all varieties of loops (positive loops and loops over negation) in normal logic programs, now complicated

---

We thank the referees for their constructive comments and suggestions. Both authors acknowledge the support from FCT/MEC NOVA LINCS PEst UID/CEC/04516/2013. Ari Saptawijaya acknowledges the support of FCT/MEC Portugal, grant SFRH/BD/72795/2010. We thank Terrance Swift and David Warren for their expert advice in dealing with implementation issues in XSB Prolog.

*Affiliated with Faculty of Computer Science at Universitas Indonesia, Depok, Indonesia.

by tabled abduction. Other challenges are pertinent to optimizations, by benefitting from XSB features, e.g., constructing dual rules by need only. Third, we evaluate TABDUAL with respect to various standpoints. The evaluations employ cases from declarative debugging, and also touch upon tabling *nogoods* of subproblems in the context of abduction.

The techniques introduced in TABDUAL intends to sensitize a general audience of users, and of implementers of various LP systems, to the potential benefits of tabled abduction, where a number of its techniques are also adaptable and importable into LP systems that afford tabling mechanisms, other than XSB Prolog.

Keywords: Abduction, Abductive Logic Programming, Dual Transformation, Tabling

1 Introduction

Abduction has been well studied in the field of computational logic, and logic programming (LP) in particular, for a few decades by now [4, 11, 14, 17, 20, 22, 44]. Abduction in LP offers a formalism to declaratively express problems in a variety of areas, e.g. in diagnosis, planning, scheduling, reasoning of rational agents, decision making, knowledge assimilation, natural language understanding, security protocols verification, and systems biology [1, 6, 16, 18, 23–25, 31, 34]. On the other hand, many Prolog systems have become mature and practical, and thus it makes sense to facilitate the use of abduction into such systems, be it two-valued abduction (as adopted in this work) or three-valued, e.g. [9].

In abduction, finding some best explanations (i.e. adequate abductive solutions) to the observed evidence, or finding assumptions that can justify a goal, can be very costly. It is often the case that abductive solutions found within one context are also relevant in a different context, and can be reused with little cost. In LP, absent of abduction, goal solution reuse is commonly addressed by employing a tabling mechanism [46]. Therefore, tabling appears to be conceptually suitable for abduction, so as to reuse abductive solutions. In practice, abductive solutions reuse is not immediately amenable to tabling, because such solutions go together with an abductive context.

In [32], we preliminarily explore the idea of how to benefit from tabling mechanisms in order to reuse priorly obtained abductive solutions, from one abductive context to another. This technique of tabling abductive solutions (*tabled abduction*) is underpinned by ABDUAL [4], a theory for computing abduction over Well-Founded Semantics. The technique presented in [32] consists of a program transformation from abductive normal logic programs into tabled logic programs, and it specifically employs the dual transformation of ABDUAL. The dual transformation allows to more efficiently handle the problem of abduction under negative goals, by introducing their positive dual counterparts.

In this paper, we formalize the tabled abduction transformation in [32], while also simplifying it, by abstracting away from implementation details, such as dealing with loops (i.e. positive loops and loops over negation) in abductive normal logic programs, non-grounded programs, etc. In other words, the present transformation focuses on an innovative re-uptake of prior abductive solution entries in tabled predicates as well as the dual transformation of ABDUAL, on which it relies.

Based on the formalization, we develop a tabled abduction system TABDUAL. The implementation of TABDUAL poses several challenges [36], both in concretely realizing the abstract theory of ABDUAL and in benefitting from features of XSB Prolog [48], in which TABDUAL is implemented, for optimizations:

1. In the theory of ABDUAL, the dual transformation does not concern itself with programs having variables. Without violating the groundness assumption in the theory of ABDUAL, the TABDUAL implementation takes care of such programs. More precisely, TABDUAL helps ground (dualized) negative subgoals and deals with non-ground negative goals. Note that, in dealing with non-ground negative goals, we look just for abductive solutions of such non-ground negative goals, and not for constraints on free variables of its calling arguments, i.e., no constructive negation is applied.

2. The tabling mechanism in XSB Prolog supports Well-Founded Semantics [51], and allows dealing with loops in the program to ensure the termination of looping queries. The implementation of TABDUAL employs XSB's tabling as much as possible to deal with loops. Nevertheless, tabled abduction introduces a complication concerning some varieties of loops. This complication is resolved by some pragmatic approaches using available tabling constructs in XSB, such as tabled negation.

3. TABDUAL allows modular mixes of abductive and non-abductive program parts, and one can benefit from the latter part by enacting a simpler translation of predicates in the program being comprised just of facts. This simpler treatment distinguishes the transformation between rules in general and predicates defined extensionally by facts alone. It particularly helps avoid superfluous transformation of facts, which would hinder the use of large factual data.

4. We address the issue of potentially heavy transformation load due to producing all dual rules in advance, regardless of their need. Such a heavy dual transformation makes it a bottleneck of the whole abduction process. A natural solution is instead to perform the dual transformation *by-need*, i.e. dual rules for a predicate are only created as their need is felt during abduction. We detail two approaches to realizing the dual transformation by-need: (a) by creating and tabling all dual rules for a predicate on the first invocation of its negation; and (b) by lazily generating and storing its dual rules in a trie (instead of tabling them), as new alternatives are required. The former

approach leads to an *eagerly by-need* dual rules tabling (under local table scheduling strategy), whereas the latter permits a *lazily by-need* dual rules construction (in lieu of batched table scheduling).

5. TABDUAL provides a system predicate that permits accessing ongoing abductive solutions. This is a useful feature and extends TABDUAL's flexibility, as it allows manipulating abductive solutions dynamically, e.g. preferring or filtering ongoing abductive solutions, e.g. checking them explicitly against nogoods at predefined program points.

TABDUAL has been evaluated with various objectives, where several TABDUAL variants (of the same underlying implementation) are examined, by separately factoring out TABDUAL's features relevant to each evaluation objective [38]. First, we evaluate the benefit of tabling abductive solutions, where we employ an example from declarative debugging to debug missing solutions of logic programs, via a process now characterized as abduction [39], instead of as belief revision [29, 30]. Second, we use the other case of declarative debugging, that of debugging incorrect solutions, to evaluate the relative worth of the dual transformation by-need. Third, we touch upon tabling so-called *nogoods* of subproblems in the context of abduction (i.e. abductive solution candidates that violate constraints), and show that tabling abductive solutions can be appropriate for tabling nogoods of subproblems. Fourth and finally, we also evaluate TABDUAL in dealing with programs having loops, where we compare its results with those obtained from an implementation of ABDUAL [5].

TABDUAL is an ongoing work, which primarily intends to sensitize a general audience of users, and of implementers of various LP systems, to the potential benefits of tabled abduction. Though TABDUAL is implemented in XSB Prolog, a number of its techniques are adaptable and importable into other LP systems that afford required tabling mechanisms. They add and aid to the considerations involved in the research of the still ongoing developments of tabling mechanisms in diverse LP systems, and serve to inspire these systems in terms of solutions, options and experimentation results of incorporating tabled abduction.

The rest of the paper is organized as follows. Section 2 provides basic definitions in LP and how abduction is accomplished in LP. Tabled abduction and the formalization of the TABDUAL transformation are presented in Section 3. We justify soundness and completeness of TABDUAL, based on the theory of ABDUAL, in Section 4, and discuss its complexity in Section 5. The aforementioned implementation aspects of TABDUAL are detailed in Section 6. Section 7 exhibits and analyses the evaluation results of TABDUAL. We conclude in Section 8, by discussing related work and further developments, including their potential joint use with other non-monotonic LP features, having their own tabling requirements and attending benefits.

2 Preliminaries

A *logic rule* has the form $H \leftarrow B_1, \ldots, B_m, not\ B_{m+1}, \ldots, not\ B_n$, where $n \geq m \geq 0$ and H, B_i with $1 \leq i \leq n$ are atoms; H and $B_1, \ldots, B_m, not\ B_{m+1}, \ldots, not\ B_n$ are called the head and the body of the rule, resp. We use '*not*' to denote default negation. The atom B_i and its default negation $not\ B_i$ are named positive and negative *literals*, resp. When $n = 0$, we say the rule is a *fact*, simply written as H. The atoms *true* and *false* are, by definition, respectively true and false in every interpretation. A rule in the form of a denial, i.e. with empty head, or equivalently with *false* as head, is an *integrity constraint* (IC). A *logic program* is a set of logic rules, where non-ground rules (i.e. rules containing variables) stand for all their ground instances. We focus on *normal logic programs*, i.e. those whose heads of rules are positive literals or empty. As usual, p/n denotes predicate p with arity n.

Abduction (inference to the best explanation – a common designation in the philosophy of science [21, 26]), is a reasoning method, whereby one chooses those hypotheses that would, if true, best explain the observed evidence (satisfy some query), while meeting any attending ICs. In LP, abductive hypotheses (*abducibles*) are named positive or negative literals of the program, which have no rules, and whose truth value is not initially assumed. Abducibles may have arguments, but for simplicity they must be ground when abduced. An *abductive normal logic program* is a normal logic program that allows abducibles appearing in the body of rules. Note that abducible '*not a*' does not refer to the default negation of abducible a, as abducibles do not appear in the head of a rule, but instead to the explicitly assumed hypothetical negation of a. The truth value of abducibles may be independently assumed *true* or *false*, via either their positive or negated form, as the case may be, to produce an abductive solution to a query, i.e. a consistent set of assumed hypotheses that support it. An *abductive solution* to a query is a consistent set of abducible instances that, when substituted by their assigned truth value everywhere in the program P, affords us with a model of P (for the specific semantics used on P), which satisfies both the query and the ICs – a so-called *abductive model*.

Abduction in LP can naturally be accomplished by a top-down query-oriented procedure to find an (abductive) solution to a query (by-need, i.e. as abducibles are encountered), where the abducibles in the solution are leaves in its procedural query-rooted call-graph, i.e. the graph recursively engendered by the procedure calls from literals in bodies of rules to heads of rules, and thence to the literals in the rule's body. This top-down computation is possible only when the underlying semantics is relevant, i.e. avoids having to computing a whole model (to guarantee its existence) in order to find an answer to a query: it suffices to use only the rules relevant to the query – those in its procedural call-graph – to find its truth value. The Well-Founded Semantics (WFS) [51] enjoys the relevance property, and thus it allows abduction to be performed by need. This is induced by the top-down query-oriented procedure, solely for finding the relevant abducibles and their truth value, whereas the values of abducibles not mentioned in the abductive solution are indifferent to the query. Tabled abduction and its prototype TABDUAL is underpinned by the theory of ABDUAL [4] that computes abduction over WFS. Note that though WFS is three-valued, the abduction mechanism in TABDUAL enforces, by design, two-valued abductive solutions; that is, needed abducibles are assumed either true or false, so as not to contribute with undefinedness towards the query.

3 Tabled abduction in TABDUAL

We start by giving the motivation for the need of tabled abduction, and subsequently show how tabled abduction is conceptualized and realized in the TABDUAL transformation.

3.1 Motivation

Example 1. *Consider an abductive logic program P_0, with a and b abducibles:*
$$q \leftarrow a. \quad s \leftarrow b, q. \quad t \leftarrow s, q.$$
Suppose three queries: q, s, and t, are individually launched, in that order. The first query, q, is satisfied simply by taking $[a]$ as the abductive solution for q, and tabling it. Executing the second query, s, amounts to satisfying the two subgoals in its body, i.e. abducing b followed by invoking q. Since q has previously been invoked, we can benefit from reusing its solution, instead of recomputing, given that the solution was tabled. That is, query s can be solved by extending the current ongoing abductive context $[b]$ of subgoal q with the already tabled abductive solution $[a]$ of q, yielding $[a, b]$. The final query t can be solved similarly. Invoking the first subgoal s results in the priorly registered abductive solution $[a, b]$, which becomes the current abductive context of the second subgoal q. Since $[a, b]$ subsumes the previously obtained (and tabled) abductive solution $[a]$ of q, we can then safely take $[a, b]$ as the abductive solution to query t. This example shows how $[a]$, as the abductive solution of the first query q, can be reused from one abductive context of q (i.e. $[b]$ in the second query, s) to its other context (i.e. $[a, b]$ in the third query, t). In practice the body of rule q may contain a huge number of subgoals, causing potentially expensive recomputation of its abductive solutions and thus such unnecessary recomputation should be avoided.

Tabled abduction in TABDUAL consists of two stages: program transformation and abduction. The program transformation produces tabled logic programs from abductive normal logic programs. Abduction is then enacted on the transformed program. Example 1 indicates two key ingredients of the transformation:

1. *abductive context*, which relays the ongoing abductive solution from one subgoal to subsequent subgoals, as well as from the head to the body of a rule, via *input* and *output* contexts, where abducibles can be envisaged as the terminals of parsing,

2. *tabled predicates*, which table the abductive solutions for predicates defined in the input program, such that they can be reused from one abductive context to another.

3.2 TABDUAL transformation

The TABDUAL transformation is underpinned by the theory of ABDUAL [4], but additionally employs the aforementioned idea of tabling and of reusing abductive solutions.

The whole TABDUAL transformation (Section 3.2.4) consists of several parts, viz., the transformations for tabling abductive solutions, for producing dualized negation, and for inserting abducibles into an abductive context. Their specifications are formalized in Sections 3.2.1, 3.2.2, and 3.2.3, respectively. Finally, queries should also be transformed, as detailed in Section 3.2.5.

3.2.1 Tabling abductive solutions

We show in Example 2, how the idea described in Example 1 can be realized by the program transformation. It illustrates how every rule in P_0 is transformed, by introducing a corresponding tabled predicate with one extra argument for the abductive solution entry, such that it can facilitate solution reuse from one abductive context to another.

Example 2. *We show first how the rule $t \leftarrow s, q$ in P_0 is transformed into two rules:*
$t_{ab}(E_2) \leftarrow s([\,], E_1), q(E_1, E_2).$ $t(I, O) \leftarrow t_{ab}(E), produce\_context(O, I, E).$
Predicate $t_{ab}(E)$ is the tabled predicate which is introduced to table one abductive solution for t in its argument E. Its definition, in the rule on the left, follows from the original definition of t. Two extra arguments, that serve as input and output contexts, are added to the subgoals s and q in the rule's body. The left rule expresses that the tabled abductive solution E_2 of t_{ab} is obtained by relaying the ongoing abductive solution stored in context E_1 from subgoal s to subgoal q in the body, given the empty input abductive context of s (because there is no abducible by itself in the body of the original rule of t). The rule on the right shows how the tabled abductive solution in E of t_{ab} can be reused for a given (input) abductive context of t. This rule expresses that the output abductive solution O of t is obtained from the solution entry E of t_{ab} and the given input context I of t, via the TABDUAL system predicate $produce\_context(O, I, E)$. This system predicate concerns itself with: whether E is already contained in I and, if not, whether there are any abducibles from E, consistent with I, that can be added to produce O. If E is inconsistent with I then the specific entry E cannot be reused with I, $produce\_context/3$ fails and another entry E is sought. In other words, $produce\_context/3$ should guarantee that it produces a consistent output context O from I and E that encompasses both.

The other two rules in P_1 are transformed following the same idea. The rule $s \leftarrow b, q$ is transformed into:

$$s_{ab}(E) \leftarrow q([b], E). \qquad s(I, O) \leftarrow s_{ab}(E), produce\_context(O, I, E).$$

where $s_{ab}(E)$ is the predicate that tables, in E, the abductive solution of s. Notice how b, the abducible appearing in the body of the original rule of s, becomes the input abductive context of q. The same transformation is obtained, even if b comes after q in the body of the rule s.

Finally, the rule $q \leftarrow a$ is transformed into:
$$q_{ab}([a]). \qquad q(I, O) \leftarrow q_{ab}(E), produce\_context(O, I, E).$$
where the original rule of q, which is defined solely by the abducible a, is simply transformed into the tabled fact $q_{ab}/1$.

Example 3. *Consider the following program that contains rules of non-nullary predicate $q/1$ with variables ($a/1$ abducible):*

$$q(0). \qquad q(s(X)) \leftarrow a(X), q(X).$$

The transformation results in rules as follows:
$$q_{ab}(0, [\,]). \qquad q_{ab}(s(X), E) \leftarrow q(X, [a(X)], E).$$
$$q(X, I, O) \leftarrow q_{ab}(X, E), produce\_context(O, I, E).$$
Notice that the single argument of $q/1$ is kept in the tabled predicate q_{ab} (as its first argument), and one extra argument is added (as its second argument) for tabling its abductive solution entry. The transformed rules $q_{ab}/2$ and $q/3$ are defined following the same idea described in Example 2.

The transformation for tabling abductive solutions is formalized in Definition 1.

Consider an abductive normal logic program P, where every integrity constraint in P with empty head is rewritten as a rule with *false* as its head, i.e. as a denial. In the sequel, we write $\bar{t}$ to denote $[t_1, \ldots, t_n]$, $n \geq 0$, and for predicate p/n, we write $p(\bar{t})$ to denote $p(t_1, \ldots, t_n)$,[1] and we write H_r and $\mathcal{B}_r$ to denote the head and the body of rule $r \in P$, respectively. Mark that abducibles do not have rules.

Definition 1 (Transformation for tabling abductive solutions). *Let $\mathcal{A}_r \subseteq \mathcal{B}_r$ be the set of abducibles (either positive or negative) in $r \in P$, and r' be the rule, such that $H_{r'} = H_r$ and $\mathcal{B}_{r'} = \mathcal{B}_r \setminus \mathcal{A}_r$.*

1. *For every rule $r \in P$ with r' the rule $l(\bar{t}) \leftarrow L_1, \ldots, L_m$, we define $\tau'(r)$:*

$$l_{ab}(\bar{t}, E_m) \leftarrow \alpha(L_1), \ldots, \alpha(L_m).$$

---

[1] In particular, we write $\bar{X}$ to denote $[X_1, \ldots, X_n]$, $p(\bar{X})$ to denote $p(X_1, \ldots, X_n)$, and $p(\bar{X}, Y, Z)$ to denote $p(X_1, \ldots, X_n, Y, Z)$, where all variables are distinct.

where α is defined as:

$$\alpha(L_i) = \begin{cases} l_i(\bar{t}_i, E_{i-1}, E_i) & , \text{if } L_i = l_i(\bar{t}_i) \\ not\_l_i(\bar{t}_i, E_{i-1}, E_i) & , \text{if } L_i = not\, l_i(\bar{t}_i) \end{cases}$$

with $1 \leq i \leq m$, E_i are fresh rule variables,[2] and $E_0 = \mathcal{A}_r$.

2. For every predicate p/n defined in P, we define $\tau^+(p)$:

$$p(\bar{X}, I, O) \leftarrow p_{ab}(\bar{X}, E), produce\_context(O, I, E).$$

where $produce\_context/3$ is a TABDUAL system predicate.

Example 4. *Consider the following program P, where rules are named with r_i and $a/1$ is an abducible.*

$$\begin{aligned} r_1: & \quad u(0,\_). \\ r_2: & \quad u(s(X), Y) \leftarrow a(X), v(X, Y, Z), not\, w(Z). \\ r_3: & \quad v(X, X, s(X)). \end{aligned}$$

We have $\mathcal{A}_{r_i}$ and r'_i, for $1 \leq i \leq 3$, as follows:[3]

- $\mathcal{A}_{r_1} = [\,]$ and $r'_1: u(0,\_).$

- $\mathcal{A}_{r_2} = [a(X)]$ and $r'_2: u(s(X), Y) \leftarrow v(X, Y, Z), not\, w(Z).$

- $\mathcal{A}_{r_3} = [\,]$ and $r'_3: v(X, X, s(X)).$

The transformation of Definition 1 results in:

$$\begin{aligned} \tau'(r_1): & \quad u_{ab}(0,\_,[\,]). \\ \tau'(r_2): & \quad u_{ab}(s(X), Y, E_2) & \leftarrow & \quad v(X, Y, Z, [a(X)], E_1), not\_w(Z, E_1, E_2). \\ \tau'(r_3): & \quad v_{ab}(X, X, s(X), [\,]). \\ \tau^+(u): & \quad u(X_1, X_2, I, O) & \leftarrow & \quad u_{ab}(X_1, X_2, E), produce\_context(O, I, E). \\ \tau^+(v): & \quad v(X_1, X_2, X_3, I, O) & \leftarrow & \quad v_{ab}(X_1, X_2, X_3, E), produce\_context(O, I, E). \end{aligned}$$

Notice that both arguments of $u/2$ are kept in the tabled predicate u_{ab} (as its first two arguments), and one extra argument is added (as its third argument) for tabling its abductive solution entry. Similar reasoning also applies to $v/3$. We do not have $\tau^+(w)$, because there is no rule of $w/1$ in the program, i.e. $w/1$ is not defined in P.

---

[2] Variables E_i serve as abductive contexts.
[3] We use Prolog list notation to represents sets.

3.2.2 Abduction under negative goals

For abducing under negative goals, the program transformation employs the *dual transformation* from ABDUAL [4]. It makes negative goals 'positive' literals, thus permitting to avoid the computation of all abductive solutions of the positive goal argument, and then having to negate their disjunction. The dual transformation enables us to obtain one abductive solution at a time, just as when we treat abduction under positive goals. The dual transformation defines for each atom A and its set of rules R in a normal program P, a set of dual rules whose head not_A is true if and only if A is false by R in the employed semantics of P. Note that, instead of having a negative goal $not\ A$ as the rules' head, we use its corresponding 'positive' literal, not_A. Example 5 illustrates the main idea of how the dual transformation is employed in TABDUAL.

Example 5. *Consider program P_2, where a is an abducible:*
$$p \leftarrow a. \qquad p \leftarrow q, not\ r. \qquad r.$$

- *With regard to p, the transformation will create a set of dual rules for p which falsify p with respect to its two rules, i.e. by falsifying both the first rule and the second rule, expressed below by predicate p^{*1} and p^{*2}, respectively:*
$$not\_p(T_0, T_2) \leftarrow p^{*1}(T_0, T_1),\ p^{*2}(T_1, T_2).$$

In the TABDUAL transformation, this single rule is known as the first layer of the dual transformation. Note the addition of the input and output abductive context arguments, T_0 and T_2, in the head, and similarly in each subgoal of the rule's body, where intermediate context T_1 relays the ongoing abductive solution from p^{*1} to p^{*2}.

The second layer contains the definitions of p^{*1} and p^{*2}, where p^{*1} and p^{*2} are defined by falsifying the body of p's first rule and second rule, respectively.

  - *In case of p^{*1}: the first rule of p is falsified only by abducing the negation of a. Therefore, we have:*
    $$p^{*1}(I, O) \leftarrow not\_a(I, O).$$
    Notice that the negation of a, i.e. $not\ a$, is abduced by invoking the subgoal $not\_a(I, O)$. This subgoal is defined via the transformation of abducibles, as discussed below.

  - *In case of p^{*2}: the second rule of p is falsified by alternatively failing one subgoal in its body at a time, i.e. by negating q or, instead, by negating $not\ r$.*
    $$p^{*2}(I, O) \leftarrow not\_q(I, O). \qquad p^{*2}(I, O) \leftarrow r(I, O).$$

- *With regard to q, the dual transformation produces the fact*
$$not\_q(I, I).$$

as its dual, because there is no rule for q in P_2. Since it is a fact, the content of the context I is simply relayed from the input to the output context, i.e. having no body, the output context does not depend on the context of any other goals, but depends only on its corresponding input context.

- *With regard to r, since it is a fact, its dual contains*

$$not\_r(T_0, T_1) \leftarrow r^{*1}(T_0, T_1).$$

but with no definition of $r^{*1}/2$. It may equivalently be defined as:

$$not\_r(\_, \_) \leftarrow fail$$

Example 5 shows that the dual rules of nullary predicates are simply defined by falsifying the bodies of their corresponding positive rules. But a goal of non-nullary predicates may also fail (or equivalently, its negation succeeds), when its arguments disagree with the arguments of its rules. For instance, if we have just a fact $q(1)$, then goal $q(0)$ will fail (or equivalently, goal $not\ q(0)$ succeeds). That is, besides falsifying the body of a rule, a dual of a non-nullary predicate can additionally be defined by disunifying its arguments and the arguments of its corresponding positive rule, as in Example 6.

Example 6. *Consider program P_5:*

$$q(0). \qquad q(s(X)) \leftarrow a(X).$$

where $a/1$ is an abducible. Let us examine the dual transformation of non-nullary predicate $q/1$.

1. $not\_q(X, T_0, T_2) \quad \leftarrow \quad q^{*1}(X, T_0, T_1), q^{*2}(X, T_1, T_2).$
2. $q^{*1}(X, I, I) \qquad\quad \leftarrow \quad X \text{ \textbackslash= } 0.$
3. $q^{*2}(X, I, I) \qquad\quad \leftarrow \quad X \text{ \textbackslash= } s(\_).$
4. $q^{*2}(s(X), I, O) \quad\ \leftarrow \quad not\_a(X, I, O).$

Line 1 shows the first layer of the dual rules for predicate $q/1$, which is defined as usual, i.e. $q/1$ is falsified by falsifying both its first and second rules. Lines 2-4 show the second layer of the dual rules for non-nullary predicates:

- In case of q^{*1}, the first rule of $q/1$, which is fact $q(0)$, is falsified by disunifying q^{*1}'s argument X with 0 (line 2). Note that, this is the only way to falsify $q(0)$, since it has no body.

- In case of q^{*2}, the second rule of $q/1$ is falsified by disunifying q^{*2}'s argument X with the term $s(\_)$ (line 3), or alternatively, by instead keeping the head unification and falsifying its body, i.e. by abducing the negation of $a/1$ (line 4).

Definition 2 provides the specification of the transformation to construct dualized negation in TABDUAL.

Definition 2 (Transformation of dualized negation). *1. For every predicate p/n, $n \geq 0$, defined in P:*

$$p(\bar{t}_1) \leftarrow L_{11}, \ldots, L_{1n_1}.$$
$$\vdots$$
$$p(\bar{t}_m) \leftarrow L_{m1}, \ldots, L_{mn_m}.$$

with $n_i \geq 0$, $1 \leq i \leq m$:

(a) *The first layer of the dual transformation is defined by $\tau^-(p)$:*

$$not\_p(\bar{X}, T_0, T_m) \leftarrow p^{*1}(\bar{X}, T_0, T_1), \ldots, p^{*m}(\bar{X}, T_{m-1}, T_m).$$

with T_i, $0 \leq i \leq m$, are fresh rule variables.[4]

(b) *The second layer of the dual transformation is defined by:*
$\tau^*(p) = \bigcup_{i=1}^{m} \tau^{*i}(p)$, and $\tau^{*i}(p)$ *is the smallest set that contains the following rules:*

$$p^{*i}(\bar{X}, I, I) \leftarrow \bar{X} \neq \bar{t}_i.$$
$$p^{*i}(\bar{t}_i, I, O) \leftarrow \sigma(L_{i1}, I, O).$$
$$\vdots$$
$$p^{*i}(\bar{t}_i, I, O) \leftarrow \sigma(L_{in_i}, I, O).$$

where σ is defined as follows:

$$\sigma(L_{ij}, I, O) = \begin{cases} l_{ij}(\bar{t}_{ij}, I, O) & , \text{if } L_{ij} = not\ l_{ij}(\bar{t}_{ij}) \\ not\_l_{ij}(\bar{t}_{ij}, I, O) & , \text{if } L_{ij} = l_{ij}(\bar{t}_{ij}) \end{cases}$$

Notice that, in case of $p/0$ (i.e. $n = 0$), rule $p^{*i}(\bar{X}, I, I) \leftarrow \bar{X} \neq \bar{t}_i$ is omitted, since both $\bar{X}$ and $\bar{t}_i$ are $[\ ]$.[5]

2. *For every predicate r/n, $n \geq 0$, in P, that has no definition, we define $\tau^-(r)$:*[6]

$$not\_r(\bar{X}, I, I).$$

---

[4] Variables T_i serve as abductive contexts.

[5] This means, when $p/0$ is defined as a fact in P, we have $not\_p(T_0, T_1) \leftarrow p^{*1}(T_0, T_1)$ in the first layer, but there is no rule of $p^{*1}/2$ in the second layer. Equivalently, it may be defined as $not\_p(\_,\_) \leftarrow fail$. (cf. the dual rule of predicate $r/0$ in Example 5).

[6] In particular, if there is no integrity constraint in P, we have $\tau^-(false) : not\_false(I, I)$.

Example 7. *Recall program P in Example 4. The transformation of Definition 2 results in:*

$\tau^-(u):$ $not\_u(X_1, X_2, T_0, T_2) \leftarrow u^{*1}(X_1, X_2, T_0, T_1), u^{*2}(X_1, X_2, T_1, T_2).$
$\tau^-(v):$ $not\_v(X_1, X_2, X_3, T_0, T_1) \leftarrow v^{*1}(X_1, X_2, X_3, T_0, T_1).$
$\tau^-(w):$ $not\_w(X, I, I).$
$\tau^-(false):$ $not\_false(X, I, I).$
$\tau^*(u):$ $u^{*1}(X_1, X_2, I, I) \leftarrow [X_1, X_2] \neq [0, \_].$
 $u^{*2}(X_1, X_2, I, I) \leftarrow [X_1, X_2] \neq [s(X), Y].$
 $u^{*2}(s(X), Y, I, O) \leftarrow not\_a(X, I, O).$
 $u^{*2}(s(X), Y, I, O) \leftarrow not\_v(X, Y, Z, I, O).$
 $u^{*2}(s(X), Y, I, O) \leftarrow w(Z, I, O).$
$\tau^*(v):$ $v^{*1}(X_1, X_2, X_3, I, I) \leftarrow [X_1, X_2, X_3] \neq [X, X, s(X)].$

3.2.3 Transforming abducibles

In Example 5, $p^{*1}(I, O)$ is defined by abducing *not a*, achieved by invoking subgoal $not\_a(I, O)$. Abduction in TABDUAL is realized by transforming each abducible atom (and its negation) into a rule, which updates the abductive context with the abducible atom (or its negation, respectively). Say, abducible a of Example 5 translates to:

$$a(I, O) \leftarrow insert\_abducible(a, I, O).$$

where $insert\_abducible(A, I, O)$ is a TABDUAL system predicate that inserts abducible A into input context I, resulting in output context O. It keeps the consistency of the context, failing if inserting A results in an inconsistent one. Abducible *not a* is transformed similarly, where *not a* is renamed into *not_a* in the head:

$$not\_a(I, O) \leftarrow insert\_abducible(not\ a, I, O).$$

The specification for the transformation of abducibles is given in Definition 3.

Definition 3 (Transformation of abducibles). *Let $\mathcal{A}_P$ be the set of abducible atoms in P. For every $a(\bar{X}) \in \mathcal{A}_P$, we define $\tau^\circ(a(\bar{X}))$ as the smallest set that contains the rules:*

$a(\bar{X}, I, O) \leftarrow insert\_abducible(a(\bar{X}), I, O).$
$not\_a(\bar{X}, I, O) \leftarrow insert\_abducible(not\ a(\bar{X}), I, O).$

where $insert\_abducible/3$ is a TABDUAL system predicate. Mark that, in the body of the second rule, 'not a' is used instead of 'not_a'.

Example 8. *Recall program P in Example 4. We have $\mathcal{A}_P = \{a(X)\}$. The transformation of Definition 3 results in:*

$\tau^\circ(a(X)):$ $a(X, I, O) \leftarrow insert\_abducible(a(X), I, O).$
 $not\_a(X, I, O) \leftarrow insert\_abducible(not\ a(X), I, O).$

3.2.4 TABDUAL program transformation

Finally, the specification of the TABDUAL program transformation is given in Definition 4.

Definition 4 (TABDUAL program transformation). *Let P be an abductive normal logic program, $\mathcal{P}_P$ be the set of predicates in P, and $\mathcal{A}_P$ be the set of abducible atoms in P. Taking:*

- $\tau'(P) = \{\tau'(r) \mid r \in P\}$
- $\tau^+(P) = \{\tau^+(p) \mid p \in \mathcal{P}_P \text{ and } p \text{ is defined}\}$
- $\tau^-(P) = \{\tau^-(p) \mid p \in \mathcal{P}_P\}$
- $\tau^*(P) = \{\tau^*(p) \mid p \in \mathcal{P}_P \text{ and } p \text{ is defined}\}$
- $\tau^\circ(P) = \{\tau^\circ(a) \mid a \in \mathcal{A}_P\}$

The TABDUAL transformation τ transforms P into $\tau(P)$, where $\tau(P)$ is defined as:

$$\tau(P) = \tau'(P) \cup \tau^+(P) \cup \tau^-(P) \cup \tau^*(P) \cup \tau^\circ(P)$$

Example 9. *The set of rules obtained in Example 4, 7, and 8 forms $\tau(P)$ of program P.*

3.2.5 Transforming queries

A query to a program, consequently, should be transformed:

- A positive goal G is simply augmented with the two extra arguments for the input and output abductive contexts.

- A negative goal $not\ G$ is made 'positive', not_G, and added the two extra input and output context arguments.

Moreover, a query should additionally ensure that all ICs are satisfied. When there is no IC defined in a program, then, following the dual transformation, fact
$$not\_false(I, I).$$
is added. Otherwise, ICs, which are rules with *false* in their heads, are transformed just like any other rules; the transformed rules with the heads $false(E)$ and $false(I, O)$ may be omitted. Finally, a query should always be conjoined with $not\_false/2$ to ensure that all integrity constraints are satisfied.

Example 10. *Query*
$$?- \; not \; p.$$
first transforms into $not\_p(I, O)$. *Then, to satisfy all ICs, it is conjoined with* $not\_false/2$, *resulting in top goal:*
$$?- \; not\_p([\,], T), \; not\_false(T, O).$$
where O is an abductive solution to the query, given initially an empty input context. Note, how the abductive solution for not_p is further constrained by passing it to the subsequent subgoal not_false for confirmation, via the intermediate context T.

Definition 5 provides the specification of the query transformation.

Definition 5 (Transformation of queries). *Let P be an abductive normal logic program and Q_P be a query to P as follows:*
$$?- \; G_1, \; \ldots, \; G_m.$$
TABDUAL *transforms query Q_P into $\Delta(Q_P)$:*
$$?- \; \delta(G_1), \; \ldots, \; \delta(G_m), not\_false(T_m, O).$$
where δ is defined as:
$$\delta(G_i) = \begin{cases} g_i(\bar{t}_i, T_{i-1}, T_i) & \text{, if } G_i = g_i(\bar{t}_i) \\ not\_g_i(\bar{t}_i, T_{i-1}, T_i) & \text{, if } G_i = not \; g_i(\bar{t}_i) \end{cases}$$

T_0 *is a given initial abductive context (or $[\,]$ by default), $1 \leq i \leq m$, T_i, O are fresh rule variables.*[7]

Example 11. *Recall program P in Example 4. Query:*
$$?- \; u(0, s(0)), \; not \; u(s(0), 0).$$
is transformed by Definition 5 into:
$$?- \; u(0, s(0), [\,], T_1), \; not\_u(s(0), 0, T_1, T_2), \; not\_false(T_2, O).$$

---

[7] Notice that O is the output abductive context, which returns the abductive solution(s) of the query.

4 Soundness and completeness

TABDUAL essentially exploits a LP engineering aspect of ABDUAL [4], by providing a self-sufficient program transform to table and reuse abductive solutions from one abductive context to another. As the TABDUAL transformation is underpinned by the theory of AB-DUAL, its soundness and completeness stem from that of ABDUAL, notably from Theorem 3.2 of [4]. This theorem is adapted below in the context of TABDUAL, whose proof details can be found therein. The definitions of abductive framework $\langle P, \mathcal{A}, I \rangle$ and abductive solution $\langle P, \mathcal{A}, \mathcal{B}, I \rangle$ are referred to Definitions 2.6 and 2.8 of [4].

Theorem 1. *Let $\langle P, \mathcal{A}, I \rangle$ be an abductive framework and $\Delta(Q_P)$ is the transform of query Q_P w.r.t. program P, following Definition 5.*

- *Soundness: If O is a solution to $\Delta(Q_P)$, then $\langle P, \mathcal{A}, O, I \rangle$ is an abductive solution for Q_P.*

- *Completeness: If $\langle P, \mathcal{A}, O, I \rangle$ is an abductive solution for Q_P, then O is a solution to $\Delta(Q_P)$.*

Note that the ABDUAL evaluation to query Q stated in Theorem 3.2 [4] is obtained by applying ABDUAL operations (cf. Definition 3.9 of [4]). With the exception of operations ABDUCTION, CO-UNFOUNDED SET REMOVAL, and ANSWER CLAUSE RESOLUTION, other operations are covered and justified by XSB's SLG operations [46]. Indeed, these SLG operations underlie tabling mechanisms employed in TABDUAL, which is implemented in XSB. Other points worthy of note to relate the theory of ABDUAL and the LP engineering aspects of TABDUAL:

- The input and output abductive contexts are operational representation of abductive context *Set* in an *abductive subgoal* $\langle L, Set \rangle$, cf. Definition 2.6 of [4].

- The system predicate *insert_abducible*/3 in Definition 3 implements the ABDUAL operation ABDUCTION. Predicate *produce_context*/3 in Definition 1 does a similar job, i.e., taking care of the union of abductive contexts that appears in ABDUAL operations ANSWER CLAUSE RESOLUTION and SIMPLIFICATION.

- In [4], two forms of dual program are introduced, i.e. folded and unfolded dual program. Their need is more theoretical, to show the soundness and completeness of the dual transformation and ABDUAL. The folded form specifically deals with infinite ground programs: it avoids infinite dual rule bodies, by swapping that infinite body possibility with a folded recurrent call, to the first body literal, followed by a folded call to the remaining body literals, and so on, possibly incurring in an infinite number of rules instead. On the other hand, the unfolded form differs from the folded one only insofar as no folding rules are defined, but is equivalent to the folded form.

Though the theory of ABDUAL underpins TABDUAL, we need not be concerned with the folded dual form in TABDUAL, as it deals only with real finite non-ground programs, whose rules stand for all their ground instances. Indeed, the dual transformation in TABDUAL of Definition 2 is another way of expressing the unfolded form. Observe that p^{*i} of Definition 2 corresponds to default conjugate $conj_D(L_{i,j_i})$ in the unfolded form, where the computation of $conj_D(L_{i,j_i})$ for each j_i is realized by the second layer of the transformation in Definition 2.

- The TABDUAL implementation caters to programs with variables and non-ground queries (cf. Example 6, and later discussed in Section 6.2). Its non-groundness does not violate the groundness assumption in the theory of ABDUAL, since one can move the head unifications of a rule to equalities in its the body, before applying the TABDUAL transformation. Recall Example 6, the two rules of $q/1$ can be rewritten as:

$$q(X) \leftarrow X = 0. \qquad q(X) \leftarrow X = s(X'),\ a(X').$$

Using the above rewritten rules, it now becomes obvious how q^{*1} and q^{*2} in Example 6 are derived by the dual transformation. Mark that, because the dual transformation needs only fail one subgoal in the body at a time, the second definition of q^{*2}, i.e. $q^{*2}(s(X), I, O) \leftarrow not\_a(X, I, O)$ is obtained by assuming the equality $X = s(X')$ in the body, but alternatively failing $a(X')$:[8]

$$q^{*2}(X, I, O) \leftarrow X = s(X'), not\_a(X', I, O).$$

which is equivalent to rule 4 in Example 6, treating back the equality $X = s(X')$ in the body as head unification modulo variable renaming. To sum up, applying the above rewriting (before the TABDUAL transformation) to rules with variables allows to avoid defining a specific dual transformation for particularly dealing with such rules. Inasmuch as head unifications of a rule are moved to equalities in its body, one can think of the ground instances of all the rules, and stick to the dual transformation of ABDUAL with its groundness assumption.

- When dealing with loops, TABDUAL relies as much as possible on XSB's tabling mechanism. This is specifically true for direct positive loops (Section 6.4.1). For positive loops in dualized negation (Section 6.4.2) and negative loops over negation

---

[8] As shown in Section 6.2.1, the implementation includes all positive literals that precede the negated literal in a dual rule. Indeed, this is just another instance of requiring one failed literal in the body and allowing to assume other (preceding) positive literals to succeed.

(Section 6.4.3) additional treatments in the implementation are required, to deal with these loops correctly, now in the presence of the TABDUAL transformation. For instance, the CWA list introduced in dealing with positive loops in dualized negation (Section 6.4.2) implements the co-unfounded set of literals and supports its corresponding operation CO-UNFOUNDED SET REMOVAL in ABDUAL. Indeed, the same technique is also implemented in the ABDUAL meta-interpreter [5]. The other treatment, viz., for negative loops over negation (Section 6.4.3), benefits from XSB's tabled negation to explicitly enact DELAYING operation in ABDUAL, now within the TABDUAL transformation. All these techniques and their rationale are detailed in Section 6.4.

In summary, constructs introduced in the TABDUAL transformation as well as the implementation technical details, such as how to deal with non-groundness and programs having loops, are just a concrete realization of the more abstract theory of ABDUAL and its operations. Other implementation aspects, e.g., the dual transformation by-need (Section 6.5), are extra optimizations pertinent to XSB features, like tabling and tries.

5 Complexity

In terms of complexity, the size of the program produced by the TABDUAL transformation is linear in the size of the input program, as shown in Theorem 2, which is similar to that of ABDUAL using the folded dual form (cf. Lemma A.5 in [4]).

Definition 6. *Let P be a finite logic program and $\mathcal{B}_r$ be the body of rule $r \in P$.*

- *$preds(P)$ denotes the number of predicates in P.*
- *$heads(P)$ denotes the number of predicates defined (i.e. with rules) in P.*
- *$rules(P)$ denotes the number of rules in P.*
- *$size(P|_p)$ denotes the size of rules in P whose head is the predicate p.*
- *$size(P)$ denotes the size of P and is defined as*

$$size(P) = \Sigma_{i=1}^{rules(P)} (1 + |\mathcal{B}_{r_i}|)$$

where $|\mathcal{B}_{r_i}|$ denotes the number of body literals in r_i.[9]

The following theorem shows that the size of the program produced by the TABDUAL transformation is linear in the size of the original program.

---
[9] That is, the size of a rule r is defined as the total number of (head and body) literals in r.

Theorem 2. *Let P be an abductive normal logic program and $\mathcal{A}_P$ be the set of abducible atoms in P. Then $size(\tau(P)) < 9.size(P) + 4.|\mathcal{A}_P|$.*

The proof is contained in Appendix A.

The problem of query evaluation to abductive frameworks is NP-complete, even for those frameworks in which entailment is based on the WFS [14]. In [4], it is shown that the complexity of an ABDUAL query evaluation is proportional to the maximal number of abducibles in any abductive subgoals, and to the number of abducible atoms in the program. In particular, if the set of abducible atoms and ICs are both empty, then the cost of query evaluation is polynomial. The complexity of TABDUAL query evaluation should naturally be based on that of ABDUAL. One may observe that the table size, used in tabling abductive solutions, would be proportional to the number of distinct (positive) subgoals in the procedural call-graph, i.e. each first call of the subgoals in a given query will table, as solution entries, the abductive solutions of the called subgoal. Besides tabling, a number of implementation aspects discussed in Section 6 may help improve performance in practice.

6 Implementation aspects of TABDUAL

Next, we discuss several aspects pertaining to the implementation of the TABDUAL transformation. The implementation aspects of TABDUAL introduced in this section aims at realizing the abstract theory of ABDUAL as well as benefitting from XSB's features for TABDUAL optimizations to foster its more practical use.

6.1 Abductive and non-abductive program parts

We start by specifying TABDUAL's input programs and its basic constructs.

Example 12. *An example of input programs of* TABDUAL:

$abds([a/1])$.
$s(X) \leftarrow prolog(atom(X)), a(X)$.
$s(X) \leftarrow prolog(nat(X)), a(X)$.

$beginProlog$.
$\quad nat(0)$. $\qquad nat(s(X)) \leftarrow nat(X)$.
$endProlog$.

The input program of TABDUAL, as shown in Example 12, may consist of two parts: abductive and non-abductive parts. Abducibles need to be declared, in the abductive part, using predicate $abds/1$, whose sole argument is the list of abducibles and their arities.

The non-abductive part is distinguished from the abductive part by the *beginProlog* and *endProlog* identifiers. Any program between these identifiers will not be transformed, i.e. it is treated as a usual Prolog program. Access to the program in the non-abductive part is established using the TABDUAL system predicate $prolog/1$. That is, $prolog/1$ is a meta-predicate that calls a user-defined predicate specified in the non-abductive part, e.g., in $prolog(nat(X))$ of Example 12. This meta-predicate's argument may also be a Prolog built-in predicate, e.g., in $prolog(atom(X))$. In essence, it executes the goal in its only argument not subject to TABDUAL's transformation.

6.2 Dealing with non-ground programs

This section touches upon abduction in programs with variables. The implementation deals with two issues, viz., grounding the dualized negative subgoals in the dual transformation and dealing with non-ground negative goals.

6.2.1 Grounding dualized negated subgoals

Example 13. *Consider program P_6, with $a/1$ abducible:*

$$q(1). \qquad r(X) \leftarrow a(X). \qquad \leftarrow q(X), r(X).$$

The TABDUAL transformation results in (notice that the last rule in P_6 is an IC):

1. $q_{ab}(1, [\,])$.
2. $q(X, I, O) \leftarrow q_{ab}(X, E), produce\_context(O, I, E)$.
3. $not\_q(X, I, O) \leftarrow q^{*1}(X, I, O)$.
4. $q^{*1}(X, I, I) \leftarrow X \backslash= 1$.

5. $r_{ab}(X, [a(X)])$.
6. $r(X, I, O) \leftarrow r_{ab}(X, E), produce\_context(O, I, E)$.
7. $not\_r(X, I, O) \leftarrow r^{*1}(X, I, O)$.
8. $r^{*1}(X, I, I) \leftarrow X \backslash= \_$.
9. $r^{*1}(X, I, O) \leftarrow not\_a(X, I, O)$.

10. $not\_false(I, O) \leftarrow false^{*1}(I, O)$.
11. $false^{*1}(I, O) \leftarrow not\_q(X, I, O)$.
12. $false^{*1}(I, O) \leftarrow not\_r(X, I, O)$.

Consider query $q(1)$, which is transformed into:

$$?-\ q(1, [\,], T), not\_false(T, O).$$

Invoking the first subgoal, $q(1, [\ ], T)$, results in $T = [\ ]$. Invoking subsequently the second subgoal, $not\_false([\ ], O)$, results in the abductive solution of the given query: $O = [not\ a(X)]$, obtained via rules 10, 12, 7, and 9. Note that rule 11, an alternative to $false^{*1}$, fails due to uninstantiated X in its subgoal $not\_q(X, I, O)$, which leads to failing rules 3 and 4. For the same reason, rule 8, an alternative to r^{*1}, also fails.

Instead of having $[not\ a(1)]$ as the abductive solution to the query $q(1)$, we have the incorrect non-ground abductive solution $[not\ a(X)]$. It does not meet our requirement, in Section 2, that abducibles must be ground on the occasion of their abduction. The problem can be remedied by instantiating X, in rule 12, thereby eventually grounding the abducible $not\ a(X)$ when it is abduced, i.e. the argument X of subgoal $not\_a/3$, in rule 9, becomes instantiated.

In the implementation, grounding a dualized negated subgoal is achieved as follows: in addition to placing a negated literal, say not_p, in the body of the second layer dual rule, all positive literals that precede literal p, in the body of the corresponding original positive rule, are also kept in the body of the dual rule. For rule 12, introducing the positive subgoal $q(X)$, originating from the positive rule, before the negated subgoal $not\_r(X, I, O)$ in the body of rule 12, helps instantiate X in this case. Rule 12 now becomes (all other rules remain the same):

$$12.\ false^{*1}(I, O) \leftarrow q(X, I, T), not\_r(X, T, O).$$

Notice that, differently from before, the rule is now defined by introducing all positive literals that appear before r in the original rule; in this case we introduce $q/3$ before $not\_r/3$. As the result, the argument X in $not\_r/3$ is instantiated to 1, due to the invocation of $q/3$, just like the case in the original rule. It eventually helps ground the the negated abducible $not\ a(X)$, when it is abduced, and the correct abductive solution $[not\ a(1)]$ to query $q(1)$ is returned. By implementing this technique, we are also able to deal with non-ground positive goals, e.g., query $q(X)$ gives the correct abductive solution as well, i.e. $[(not\ a(1)]$ for $X = 1$.

There are some points to remark on regarding this implementation technique. First, the semantics of dual rules does not change because the conditions for failure of their positive counterpart rules are that one literal must fail, even if the others succeed. The cases where the others do not succeed are handled in the other alternatives of dual rules. Second, this technique may benefit from the TABDUAL's tabled predicate, e.g. q_{ab} for predicate q, as it helps avoid redundant derivations of the newly introduced positive literals in dual rules. Finally, knowledge of shared variables in the body and whether they are local or not, may be useful to avoid introducing positive literals that are not contributing to further grounding.

6.2.2 Non-ground negative goals

Example 14. *Consider program P_7, with $a/1$ abducible:*

$$p(1) \leftarrow a(1). \qquad p(2) \leftarrow a(2).$$

Query $p(X)$ to program P_7 succeeds under TABDUAL, giving two abductive solutions: $[a(1)]$ and $[a(2)]$ for $X = 1$ and $X = 2$, respectively. But query $not\ p(X)$ does not deliver the expected solution. Instead of returning the abductive solution $[not\ a(1), not\ a(2)]$ for any instantiation of X, it returns $[not\ a(1)]$ for a particular $X = 1$. In order to find the culprit, we first look into the definition of $not\_p/3$:

1. $not\_p(X, I, O) \leftarrow p^{*1}(X, I, T), p^{*2}(X, T, O).$
2. $p^{*1}(X, I, I) \leftarrow X \mathbin{\backslash}= 1.$
3. $p^{*1}(1, I, O) \leftarrow not\_a(1, I, O).$
4. $p^{*2}(X, I, I) \leftarrow X \mathbin{\backslash}= 2.$
5. $p^{*2}(2, I, O) \leftarrow not\_a(2, I, O).$

Recall that query $not\ p(X)$ is transformed into:

$$?-\ not\_p(X, [\,], N), not\_false(N, O).$$

When the goal $not\_p(X, [\,], N)$ is launched, it first invokes $p^{*1}(X, [\,], T)$. It succeeds by the second rule of p^{*1}, in line 3 (the first rule, in line 2, fails it), with variable X is instantiated to 1 and T to $[not\ a(1)]$. The second subgoal of $not\_p(X, [\,], N)$ is subsequently invoked with the same instantiation of X and T, i.e. $p^{*2}(1, [not\ a(1)], O)$, and it succeeds by the first rule of p^{*2}, in line 4, and results in $N = [not\ a(1)]$. Since there is no IC in P_6, the abductive solution $[not\ a(1)]$ is just relayed from N to O, due to the fact $not\_false(I, I)$ in the transformed program (cf. Section 3.2.5), thus returning the abductive solution $[not\ a(1)]$ with $X = 1$ for the given query.

The culprit of this wrong solution is that both subgoals of $not\_p/3$, i.e. $p^{*1}/3$ and $p^{*2}/3$, share the argument X of $p/1$. This should not be the case, as $p^{*1}/3$ and $p^{*2}/3$ are derived from two different rules of $p/1$, hence failing p should be achieved by invoking p^{*1} and p^{*2} with an independent argument X. In other words, different variants of the calling argument X should be used in $p^{*1}/3$ and $p^{*2}/3$, as shown for rule $not\_p/3$ (line 1) below:

1. $not\_p(X, T_0, T_2) \leftarrow copy\_term([X], [X_1]), p^{*1}(X_1, T_0, T_1),$
   $copy\_term([X], [X_2]), p^{*2}(X_2, T_1, T_2).$

where the Prolog built-in predicate $copy\_term/2$ provides a variant of the list of arguments; in this example, we simply have only one argument, i.e. $[X]$.

Now, $p^{*1}/3$ and $p^{*2}/3$ are invoked using variant independent calling arguments, viz., X_1 and X_2, respectively. The same query first invokes $p^{*1}(X_1, [\,], T_1)$, which results in $X_1 = 1$ and $T_1 = [not\ a(1)]$ (by the second rule of p^{*1}), and subsequently invokes $p^{*2}(X_2, [not\ a(1)], T_2)$, resulting in $X_2 = 2$ and $T_2 = [not\ a(1), not\ a(2)]$ (by the second rule of p^{*2}). It eventually ends up with the expected abductive solution: $[not\ a(1), not\ a(2)]$ for any instantiation of X, i.e. X remains unbound.

The technique ensures, as this example shows, that $p(X)$ fails for every X, and its negation, $not\ p(X)$, hence succeeds. The dual rules produced for the negation are tailored to be, by definition, an 'if and only if' with regard to their corresponding positive rules. If we added the fact $p(Y)$ to P_7, then the same query $not\ p(X)$ would not succeed because now we have the first layer dual rule:

$$\begin{aligned}
not\_p(X, T_0, T_3) \leftarrow\ & copy\_term([X], [X_1]), p^{*1}(X_1, T_0, T_1),\\
& copy\_term([X], [X_2]), p^{*2}(X_2, T_1, T_2),\\
& copy\_term([X], [X_3]), p^{*3}(X_3, T_2, T_3).
\end{aligned}$$

and an additional second layer dual rule $p^{*3}(X, \_, \_) \leftarrow X \neq \_$ that always fails; its abductive contexts are thus irrelevant.

6.3 Transforming predicates with facts only

TABDUAL transforms predicates that comprise of just facts as any other rules in the program (cf. fact $q(1)$ and its transformed rules, in Example 13). This is clearly superfluous as facts do not induce any abduction, and the transformation would be unnecessarily heavy for programs with large factual data, which is often the case in many real world problems.

A predicate, say $q/1$, comprised of just facts, can be much more simply transformed. The transformed rules $q_{ab}/2$ and $q/3$ can be substituted by a single rule:
$$q(X, I, I) \leftarrow q(X).$$
and their negations, rather than using dual rules, can be transformed to a single rule:
$$not\_q(X, I, I) \leftarrow not\ q(X).$$
independently of the number of facts $q/1$ are there in the program. Note that the input and output context arguments are added in the head, and the input context is just passed intact to the output context. Both rules simply execute the fact calls.

Facts of predicate $q/1$ can thus be defined in the non-abductive part of the input program. For instance, if a program contains facts $q(1)$, $q(2)$, and $q(3)$, they are listed as:
$$beginProlog. \quad q(1). \quad q(2). \quad q(3). \quad endProlog.$$

Though this new transformation for facts seems trivial, it considerably improves the performance, in particular if we deal with abductive logic programs having large factual data. In this case, not only the time and space in the transformation stage can be reduced, but also in the abduction.

6.4 Dealing with loops

The tabling mechanism in XSB supports the Well-Founded Semantics, therefore it allows dealing with loops in the program, ensuring termination of looping queries. In TABDUAL, tabling for loops is taken care by its transformation, as shown in this section. TABDUAL relies as much as possible on XSB's tabling mechanism in dealing with loops; this is specifically the case for direct positive loops (Section 6.4.1). Nevertheless, the presence of tabled abduction requires some varieties of loops, viz., positive loops in (dualized) negation and negative loops over negation, to be handled carefully in the transformation, as we detail in Sections 6.4.2 and 6.4.3, respectively. Additional examples, besides the ones below, are to be found in Appendix B.

6.4.1 Direct positive loops

Example 15. *Consider program P_8 which involves a direct positive loop between predicates:*

$$p \leftarrow q. \qquad q \leftarrow p.$$

The tabling mechanism in XSB would detect direct positive loops and fail predicates involved in such loops. The TABDUAL transformation may simply benefit from it. For P_8, query p fails, due to the direct positive loop between tabled predicates p_{ab} and q_{ab}:

$$p_{ab}(E) \leftarrow q([\,], E). \qquad p(I, O) \leftarrow p_{ab}(E), produce\_context(O, I, E).$$
$$q_{ab}(E) \leftarrow p([\,], E). \qquad q(I, O) \leftarrow q_{ab}(E), produce\_context(O, I, E).$$

On the other hand, query $not\ p$ should succeed with the abductive solution: [ ]. But, instead of succeeding, this query will loop indefinitely! Recall that the call to query $not\ p$, after the transformation, becomes $not\_p([\,], T), not\_false(T, O)$. The indefinite loop occurs in $not\_p([\,], T)$ because of the mutual dependency between not_p and not_q through p^{*1} and q^{*1}:

$$not\_p(I, O) \leftarrow p^{*1}(I, O). \qquad p^{*1}(I, O) \leftarrow not\_q(I, O).$$
$$not\_q(I, O) \leftarrow q^{*1}(I, O). \qquad q^{*1}(I, O) \leftarrow not\_p(I, O).$$

The dependency creates a positive loop on negative non-tabled predicates, and such loops should succeed, precisely because the corresponding source program's loop is a direct one on positive literals, which hence must fail. We now turn to how to deal with such loops in TABDUAL.

6.4.2 Positive loops in (dualized) negation

Since any source program's direct positive loops must fail, the loops between their corresponding transformed negations, i.e. positive loops in dualized negation (introduced via the dual transformation), must succeed [4]. For instance, whereas $r \leftarrow r$ fails query r, perforce

$not\_r \leftarrow not\_r$ succeeds query not_r. This is formalized in the notion of co-unfounded set of literals in ABDUAL (cf. Definition 3.5 of [4])

We detect positive loops in (dualized) negation, PLoN for short, by tracking the ancestors of negative subgoals, whenever they are called from other negative subgoals. In the transformation, a list of ancestors, dubbed the *close-world-assumption (CWA) list* is maintained. It contains only negative literals and serves as another extra argument in the first and second layers of dual rules. Indeed, the role of this ancestor list to deal with PLoN implements the co-unfounded set of literals in ABDUAL.

The TABDUAL transformation, with PLoN detection, of P_8 results in the following first and second layers of dual rules (other transformed rules remain the same):

1. $not\_p(I, I, C) \leftarrow member(not\ p, C), !.$
2. $not\_p(I, O, C) \leftarrow p^{*1}(I, O, C).$
3. $p^{*1}(I, O, C) \leftarrow not\_q(I, O, [not\ p \mid C]).$

4. $not\_q(I, I, C) \leftarrow member(not\ q, C), !.$
5. $not\_q(I, O, C) \leftarrow q^{*1}(I, O, C).$
6. $q^{*1}(I, O, C) \leftarrow not\_p(I, O, [not\ q \mid C]).$

The CWA list C is only updated in the second layer of dual rules (cf. rules p^{*1} and q^{*1} in line 3 and 6, respectively), i.e. by adding the negative literal corresponding to the dual rule into list C. For example, in case of p^{*1} (line 3), $not\ p$ is added into the CWA list C. Note that, since the CWA list is intended to detect PLoN, the list is reset in positive subgoals occurring in the body of a dual rule. This guarantees that there are no interposing positive calls between the negative calls and their ancestor, which would break such loops.

The updated CWA list C is then used to detect PLoN via an additional rule of not_p (line 1, and similarly in line 4, for not_q). The idea is to test, whether we are returning to the same call of not_p, which is simply realized by a membership testing. If that is the case, the output context is set equal to the input context, and PLoN is anticipated by immediately succeeding not_p with the extra cut to prevent the call to the next not_p rule (which would otherwise lead to looping). This detection technique thus establishes the CO-UNFOUNDED SET REMOVAL operation in the ABDUAL theory.

This technique of PLoN detection consequently requires query $not\ p$ to be transformed into:

$$?-\ not\_p([\,], T, [\,]), not\_false(T, O).$$

i.e. it is initially called with an empty CWA list.

6.4.3 Negative loops over negation

The other type of loops that XSB's tabling mechanism already properly deals with, is the negative loops over negation (NLoN).

Example 16. *Consider program P_9:*

$$p \leftarrow q. \qquad q \leftarrow not\ p.$$

In XSB, the tabling mechanism makes p and q (also their default negations) undefined. But under TABDUAL, query p (also q) will fail, instead of being undefined. It fails, because the tabled predicate p_ab is involved in a direct positive loop as shown in the transformation below:

$$\begin{aligned}
p(I,O) &\leftarrow p_{ab}(E), produce\_context(O,I,E). \\
p_{ab}(E) &\leftarrow q([\,],E). \\
q(I,O) &\leftarrow q_{ab}(E), produce\_context(O,I,E). \\
q_{ab}(E) &\leftarrow not\_p([\,],E). \\
\\
not\_p(I,O) &\leftarrow p^{*1}(I,O). \\
p^{*1}(I,O) &\leftarrow not\_q(I,O). \\
not\_q(I,O) &\leftarrow q^{*1}(I,O). \\
q^{*1}(I,O) &\leftarrow p(I,O).
\end{aligned}$$

More precisely, whereas in the original program P_9, q is defined by the negative subgoal *not p*, in the resulting transformation q is defined by the positive subgoal *not_p* via the tabled predicate q_{ab}.

One way to resolve the problem is to wrap the positive subgoal *not_p* in the body of the rule q_{ab} with the tabled negation predicate (*tnot*/1 in XSB) twice: on the one hand it preserves the semantics of the rule (keeping the truth value by applying *tnot* twice), and on the other hand introducing *tnot* creates NLoN (instead of direct positive loops). Predicate q_{ab} is thus defined as follows (other transformed rules remain the same):

1. $q_{ab}(E) \leftarrow not\_p_{tu}([\,],E).$

2. $not\_p_{tu}(I,I) \leftarrow call\_tv(tnot\ over(not\_p(I)), undefined).$
3. $not\_p_{tu}(I,O) \leftarrow call\_tv(tnot\ over(not\_p(I)), true), p^{*1}(I,O).$

4. $not\_p(I) \leftarrow p^{*1}(I,\_).$

Here, $tnot\ over(not\_p(I))$ is the double-wrapping of *not_p* with *tnot*. It is realized via the intermediate tabled predicate *over*/1, defined as:

$$over(G) \leftarrow tnot(G).$$

The double-wrapping is called through a new auxiliary predicate $not\_p_{tu}/2$. The XSB system predicate $call\_tv/2$ calls the double-wrapping and is used to distinguish the two cases (lines 2 and 3): whether NLoN exists or not. In the former case, the returned truth value is undefined; therefore not_p_{tu} itself is undefined and its input context is simply relayed to the output context. In the latter case, where the returned truth value is true, the output context O of not_p_{tu} is obtained from the input context I as usual, i.e. by invoking $p^{*1}(I, O)$.

Notice that, instead of using the existing $not\_p(I, O)$ in the double-wrapping, we use an auxiliary predicate $not\_p(I)$ to avoid floundering in the call to $over/1$, due to the uninstantiated output context O. For this reason, the newly introduced $not\_p/1$ is thus free from the output context, but otherwise defined exactly as $not\_p/2$.

6.5 Dual transformation by-need

The TABDUAL transformation conceptually constructs *all* (first and second layer) dual rules, in advance, for every defined atom in an input program, regardless whether they are needed in abduction. We refer to this conceptual construction of dual rules in the sequel as the dual transformation STANDARD.

The dual transformation STANDARD should be avoided in practice, as potentially large sets of dual rules are created in the transformation, though only a few of them might be invoked during abduction. As real world problems typically consist of a huge number of rules, such dual transformation may suffer from a heavy computational load, and therefore hinders the subsequent abduction phase to take place, not to mention the compile time, and space requirements, of the large thus produced transformed program.

One solution to this problem is to compute dual rules *by-need*. That is, dual rules are concretely created in the abduction stage (rather than in the transformation stage), based on the need of the on-going invoked goals. The transformed program still contains the single first layer rule of the dual transformation, but its second layer is defined using a newly introduced TABDUAL system predicate, which will be interpreted by the TABDUAL system on-the-fly, during abduction, to produce the concrete rule definitions of the second layer.

Example 17. *Recall Example 5. The dual transformation by-need contains the same first layer:* $not\_p(T_0, T_2) \leftarrow p^{*1}(T_0, T_1), p^{*2}(T_1, T_2)$. *But the second now contains, for each* $i \in \{1, 2\}$:

$$p^{*i}(I, O) \leftarrow dual(i, p, I, O).$$

Predicate $dual/4$ is a TABDUAL system predicate, introduced to facilitate the dual transformation by-need:

1. It constructs generic dual rules, i.e. dual rules without any context attached to them, by-need, from the i-th rule of $p/1$, during abduction,

2. It instantiates the generic dual rules with the provided arguments and input context,

3. Finally, it subsequently invokes the instantiated dual rules.

The dual transformation by-need clearly reduces the size of the second layer dual rules. Recall from Definition 6, that the size of a rule r is the total number of (head and body) literals in p. Therefore, each rule r^{*i} has a constant size 4, which is obtained from its two rules: (1) a rule that disunifies its arguments, viz. $r^{*i}(\bar{X}, I, I) \leftarrow \bar{X} \neq \bar{t}_i$ and, (2) a rule defined with the above TABDUAL system predicate $dual/4$, viz. $r^{*i}(\bar{t}_i, I, O) \leftarrow dual(i, r(\bar{t}_i), I, O)$. Compared to the dual transformation STANDARD, whose size of $\tau^*(P)$ for constructing its second layer dual rules $size(\tau^*(P)) = 2.size(P)$ (cf. proof of Theorem 1, in Appendix A), the size of $\tau^*_{need}(P)$ for the dual transformation by-need, assuming there are m rules of r, is $size(\tau^*_{need}(P)) = \Sigma_{i=1}^{heads(P)} 4m = 4.rules(P)$. Given the definition of $size(P)$ in Definition 6, it is straightforward to verify, that the advantage of the dual transformation by-need over the STANDARD one becomes apparent when the body of r contains as little as two literals, rendering $size(\tau^*_{need}(P)) < size(\tau^*(P))$. More concrete evaluation of the dual-transformation by-need is exemplified and discussed in Section 7.3.

Having said that, constructing dual rules on-the-fly clearly introduces some extra cost in the abduction stage. Such extra cost can be reduced by memoizing the already constructed generic dual rules. Therefore, when such dual rules are later needed, they are available for reuse and their recomputation avoided.

We examine two approaches for memoizing generic dual rules for the dual transformation by-need. Their definitions of system predicate $dual/4$ are different, which distinguish how generic dual rules are constructed by-need. The first approach (Section 6.5.1) benefits from tabling to memoize generic dual rules, whereas the second one (Section 6.5.2) employs XSB's trie data structure [50]. They are referred in the sequel as BY-NEED(EAGER) and BY-NEED(LAZY), respectively, due to their dual rules construction mechanisms.

6.5.1 Dualization BY-NEED(EAGER): tabling generic dual rules

The straightforward choice for memoizing generic dual rules is to use tabling. The system predicate $dual/4$ is defined as follows (abstracting away irrelevant details):

$$dual(N, P, I, O) \leftarrow dual\_rule(N, P, Dual), call\_dual(P, I, O, Dual).$$

where $dual\_rule/3$ is a *tabled* predicate that constructs a generic dual rule $Dual$ from the N-th rule of atom P, and $call\_dual/4$ instantiates $Dual$ with the provided arguments of P

and the input context I. It also invokes the instantiated dual rule to produce the abductive solution in O.

Though predicate $dual/4$ helps realize the construction of dual rules by-need, i.e. only when a particular p^{*i} is invoked, this approach results in the *eager* construction of all dual rules for the i-th rule of predicate p, because of tabling (assuming XSB's local table scheduling is in place, rather than its alternative, in general less efficient, batched scheduling). For instance, in Example 5, when $p^{*2}(I, O)$ is invoked, which subsequently invokes $dual\_rule(2, p, Dual)$, all two alternatives of dual rules from the second rule of p, i.e. $p^{*2}(I, O) \leftarrow not\_q(I, O)$ and $p^{*2}(I, O) \leftarrow r(I, O)$ are constructed before $call\_dual/4$ is invoked for each of them. This is a bit against the spirit of a full by-need dual transformation, where only one alternative dual rule is constructed at a time, just before it is invoked. That is, generic dual rules could be constructed *lazily*.

As mentioned earlier, the reason behind this eager by-need construction is the local table scheduling strategy, that is employed by default in XSB. This scheduling strategy may not return any answers out of a strongly connected component (SCC) in the subgoal dependency graph, until that SCC is completely evaluated [48].

Alternatively, batched scheduling is also implemented in XSB, which allows returning answers outside of a maximal SCC as they are derived. In terms of the dual rules construction by-need, this means $dual\_rule/3$ would allow dual rules to be lazily constructed. That is, only one generic dual rule is produced at a time before it is instantiated and invoked. Since the choice between the two scheduling strategies can only be made for the whole XSB installation, and is not (as yet) predicate switchable, we pursue another approach to implement lazy dual rule construction.

6.5.2 Dualization BY-NEED(LAZY): storing generic dual rules in a trie

Trie is a tree data structure that allows data, such as strings, to be compactly stored by a shared representation of their prefixes. That is, all the descendants of a node in a trie have a common prefix of the string associated with that node.

XSB offers a mechanism for facts to be directly stored and manipulated in tries. Figure 1, taken from [50], depicts a trie that stores a set of Prolog facts:

$$\{rt(a, f(a, b), a), rt(a, f(a, X), Y), rt(b, V, d)\}.$$

For trie-dynamic code, trie storage has advantages, both in terms of space and time [50]:

- A trie can use much less space to store many sets of facts than standard dynamic code, as there is no distinction between the index and the code itself.

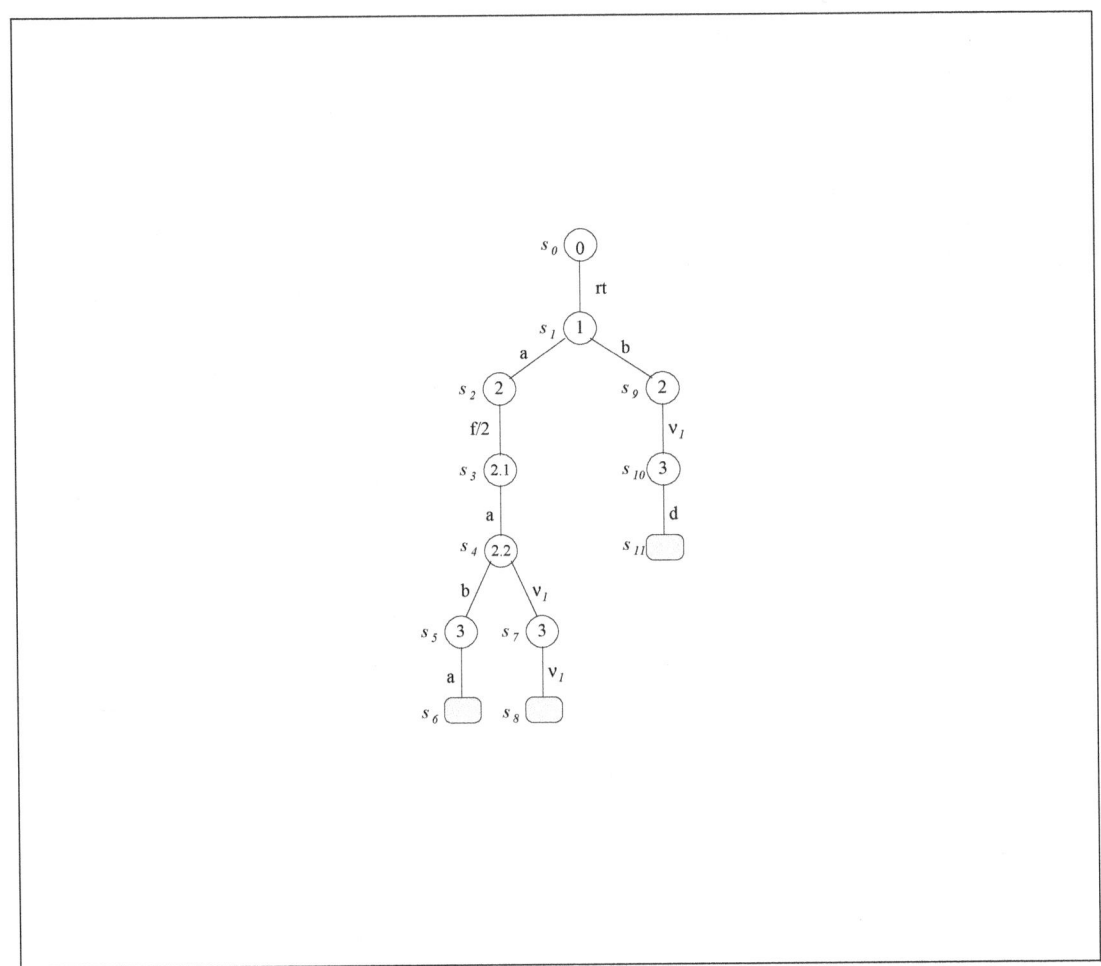

Figure 1: Facts stored as a trie.

- Directly inserting into or deleting from a trie is faster (up to 4-5 times) than with standard dynamic code, as discrimination can be made on a position anywhere in a fact.

XSB provides predicates for inserting terms into a trie, unifying a term with terms in a trie, and other trie manipulation predicates, both in the low-level and high-level API.

Generic dual rules can be represented as facts, thus once they are constructed, they can be memoized and later (a copy) retrieved and reused. Given the aforementioned advantages for storing dynamic facts and XSB support for its manipulation, a trie is preferable to the common Prolog database to store dynamically generated (i.e., by-need) dual rules. The availability of XSB system predicates to manipulate terms in a trie permits explicit control in lazily constructing generic dual rules compared to the more eager tabling approach (Section 6.5.1), as detailed below.

A fact of the form $d(N, P, Dual, Pos)$ is used to represent a generic dual rule $Dual$ from the N-th rule of P with the additional tracking information Pos, which informs the position of the literal used in constructing each dual rule. In the current TABDUAL implementation, we opt for the low-level API trie manipulation predicates, as they can be faster than the higher-level API.

Using this approach, the system predicate $dual/4$ is defined as follows (abstracting away irrelevant details):

1. $dual(N, P, I, O) \leftarrow trie\_property(T, alias(dual)), dual(T, N, P, I, O).$

2. $dual(T, N, P, I, O) \leftarrow trie\_interned(d(N, P, Dual, \_), T),$
   $call\_dual(P, I, O, Dual).$

3. $dual(T, N, P, I, O) \leftarrow current\_pos(T, N, P, Pos),$
   $dualize(Pos, Dual, NextPos),$
   $store\_dual(T, N, P, Dual, NextPos),$
   $call\_dual(P, I, O, Dual).$

Assuming that a trie T with alias $dual$ has been created, predicate $dual/4$ (line 1) is defined by an auxiliary predicate $dual/5$ with an access to the trie T, the access being provided by the trie manipulation predicate $trie\_property/2$. Lines 2 and 3 give the definition of $dual/5$. In the first definition (line 2), an attempt is made to reuse generic dual rules, which are stored already as facts $d/4$ in trie T. This is accomplished by unifying terms in T with $d(N, P, Dual, \_)$, one at a time through backtracking, via the trie manipulation predicate $trie\_interned/2$. Predicate $call\_dual/4$ then does the job as before. The second definition (line 3) constructs generic dual rules lazily. It finds, via $current\_pos/4$, the current position Pos of the literal from the N-th rule of P, which can be obtained from the last argument of fact $d(N, P, Dual, Pos)$ stored in trie T. Using this Pos information, a new generic dual rule $Dual$ is constructed by means of $dualize/3$. The latter predicate additionally updates the position of the literal, $NextPos$, for the next dualization. The dual rule $Dual$, together with the tracking information, is then memoized as a fact $d(N, P, Dual, NextPos)$ in trie T, via $store\_dual/5$. Finally, the just constructed dual $Dual$ is instantiated and invoked using $call\_dual/4$.

Whereas the first approach constructs generic dual rules by-need eagerly, the second one does it lazily. But this requires memoizing dual rules to be carried out explicitly, and additional tracking information is needed to correctly pick up on dual rule generation at the point where it was last left. This approach affords us a simulation of batched table scheduling for $dual/5$, within the default local table scheduling.

6.6 Accessing ongoing abductive solutions

TABDUAL encapsulates the ongoing abductive solution in an abductive context, which is relayed from one subgoal to another. In many problems, it is often the case that one needs to access the ongoing abductive solution in order to manipulate it dynamically, e.g. to filter abductive solutions using preferences, or eliminate so-called *nogood* combinations (those known to violate constraints). But since it is encapsulated in an abductive context, and such a context is only introduced in the transformed program, the only way to accomplish it would be to modify directly the transformed program rather than the original problem representation. This is inconvenient and clearly unpractical when we deal with real world problems with a huge number of rules.

The aforementioned issue is overcome by introducing the TABDUAL system predicate $process\_ongoing(P)$ that allows to access the ongoing abductive solution and to manipulate it, while also allowing to abduce further, using the rules of P. This system predicate is transformed by unwrapping it and adding an extra argument to P (besides the usual input and output context arguments) for the ongoing abductive solution.

Example 18. *Consider a fragment of an input program:*

$$q \leftarrow r, process\_ongoing(s). \qquad s(X) \leftarrow v(X).$$

Notice that, predicate s wrapped by $process\_ongoing/1$ has no argument; more precisely, one less argument than its definition, i.e. rule s on the right. The extra argument of rule s is indeed dedicated for the ongoing abductive solution. The tabled predicate q_{ab} in the transformed program is defined as follows:

$$q_{ab}(E) \leftarrow r([\,], T), s(T, T, E).$$

That is, $s/3$ now gets access to the ongoing abductive solution T from $r/2$, via its additional first argument. It still has the usual input and output contexts, T and E, respectively, in its second and third arguments. It indicates that, while manipulating the ongoing abduction solution, abduction may take place in s. Rule $s/1$ transforms as usual.

The system predicate $process\_ongoing/1$ permits modular mixes of abductive and non-abductive program parts. For instance, the rule of $s/1$ in P_3 may be defined by some predicates from the non-abductive program part, e.g. the rule of $s/1$ can be defined instead as:

$$s(X) \leftarrow prolog(preferred(X)), a(X).$$

where $a/1$ is an abducible and $preferred(X)$ defines, in the non-abductive program part, some preference rule on a given solution X.

6.7 Other implementation aspects

Various other aspects have also been considered in implementing TABDUAL:

- XSB's built-in predicate $numbervars/1$ is used to help writing variables, e.g. arguments of a predicate, in transformed programs. This is to avoid the problem of mixing of variables writing due to stack expansion (or garbage collection), a bug that occurs in most Prolog systems [49]. This problem particularly arises when we deal with rather big input programs.

- The list of abductive solutions is represented using two separate lists: the lists of positive and negative parts. This enables faster consistency checking of abductive solutions, in predicates $insert\_abducible/3$ and $produce\_context/3$. That is, to check consistency with respect to a literal, only the list of literals with different polarity is inspected; there is no need to traverse all literals. Moreover, both lists are ordered, in order to improve efficiency.

- The second layer dual rules are defined by giving priority to abducibles. For instance, given rule $p \leftarrow q, a$ (where a is an abducible), the first rule for p^{*1} will be $p^{*1} \leftarrow not\_a$, instead of $p^{*1} \leftarrow not\_q$ (even though, in the body of the corresponding positive rule, a comes later than q). In this way, it allows obtaining abductive solutions to negative goals earlier: not_a is returned first before not_q is invoked (the latter could involve a deep derivation before it successfully abduces a solution). Also, since the abducible will be required anyway, giving it priority may constrain earlier any solutions. Of course, care has to be taken when we deal with rules having variables, in particular concerning grounding issues (cf. Section 6.2.1). Knowledge of shared variables in the body, and whether they are local or not, may help in this case. Furthermore, the use of a domain predicate for abducibles may come in handy.

- When a program contains NLoN, the dual rules of some predicates are also tabled. These are the predicates that appear as negative subgoals in the bodies of rules. Recall the definition of q_{ab}, in Section 6.4.3, where rules not_p_{tu} are introduced for the negative goal $not\ p$ that appears in the body of rule q. Predicate not_p_{tu} is in turn defined by $not\_p/1$; the latter predicate is defined by invoking the dual rules of p: in that example, $p^{*1}/2$ (line 4). By tabling $p^{*1}/2$, its recomputation, when it is subsequently invoked as the last subgoal of the not_p_{tu}'s second rule (line 3), can be avoided.

7 Evaluation of TABDUAL

We evaluate TABDUAL from various standpoints:

1. Our first evaluation (Section 7.2) aims at evaluating the benefit of tabling abductive solutions.

2. We evaluate in Section 7.3 the relative worth of three approaches of the dual transformation (as discussed in Section 6.5), viz., STANDARD, BY-NEED(EAGER), BY-NEED(LAZY).

3. We touch upon the evaluation of tabling nogoods of subproblems in abduction, in Section 7.4.

4. Finally, in Section 7.5 we evaluate TABDUAL, implementing the technique discussed in Section 6.4, for programs with loops, and compare the results with those returned by the ABDUAL meta-interpreter of [5].

In the first two evaluations, examples from declarative debugging are employed, which are now characterized as abduction [39], rather than as belief revision [29, 30]. We specifically consider two cases of declarative debugging: missing solutions and incorrect solutions, in the evaluation of Sections 7.2 and 7.3, respectively. Before discussing the aforementioned four evaluations, we revisit in Section 7.1 our declarative debugging of definite logic programs viewed as abduction. For normal logic programs, the reader is referred to [39].

7.1 Declarative debugging of definite logic programs

Example 19. *Consider program P_{10} [29] as a buggy program:*

$$a(1). \quad a(X) \leftarrow b(X), c(Y,Y).$$
$$b(2). \quad b(3). \quad c(1,X). \quad c(2,2).$$

7.1.1 Incorrect solutions

Suppose that $a(3)$ is an incorrect solution. To debug its cause, the program is first processed using the transformation in [30], by adding default literal $not\ incorrect(i, [X_1, \ldots, X_n])$ to the body of each i-th rule of P_{10}, to defeasibly assume their correctness by default, where n is the rule's arity and X_is, for $1 \leq i \leq n$, its head arguments. This yields program P'_{10}:

$a(1) \leftarrow not\ incorrect(1, [1]).$ $\qquad a(X) \leftarrow b(X), c(Y,Y), not\ incorrect(2, [X]).$
$b(2) \leftarrow not\ incorrect(3, [2]).$ $\qquad b(3) \leftarrow not\ incorrect(4, [3]).$
$c(1, X) \leftarrow not\ incorrect(5, [1, X]).$ $\qquad c(2, 2) \leftarrow not\ incorrect(6, [2, 2]).$

In terms of abduction, one can envisage $incorrect/2$ as an abducible. To express, while debugging, that $a(3)$ is an incorrect solution, we add to P'_{10} an IC: $\leftarrow a(3)$. Running

TABDUAL on P'_{10} returns three solutions as the possible sufficient causes of the incorrect solution:

$$[incorrect(2, [3])], [incorrect(4, [3])], [incorrect(5, [1, 1]), incorrect(6, [2, 2])].$$

7.1.2 Missing solutions

Suppose $a(5)$ should be a solution of P_{10}, but is missing. To find this bug, P_{10} is transformed [29] by adding to each predicate p/n a rule:

$$p(X_1, \ldots, X_n) \leftarrow missing(p(X_1, \ldots, X_n)).$$

That is, P_{10} is transformed into P''_{10} that contains all rules from P_{10} plus three new rules:

$$a(X) \leftarrow missing(a(X)). \quad b(X) \leftarrow missing(b(X)). \quad c(X, Y) \leftarrow missing(c(X, Y)).$$

Similarly to before, $missing/1$ can be viewed as an abducible. But now, to express that we miss $a(5)$ as a solution, we add to P''_{10} an IC: $\leftarrow not\ a(5)$. TABDUAL returns the three abductive solutions on P''_{10} as the causes of missing solution $a(5)$ in P_{10}:

$$[missing(a(5))], [missing(b(5))], [missing(b(5)), missing(c(X, X))].$$

Differently from [29, 30], where minimal solutions are targeted, TABDUAL also returns non-minimal solution $[missing(b(5)), missing(c(X, X))]$. Finding minimal abductive solutions is not always desired – here bugs may well not be minimal and, in this case, TABDUAL allows one to identify and choose those bugs that satisfice so far, and to continue searching for more solutions if needed.

Next, we discuss the evaluation of TABDUAL. The experiments in all evaluations were run under XSB-Prolog 3.3.7 on a 2.26 GHz Intel Core 2 Duo with 2 GB RAM. The time indicated in all results refers to the CPU time (as an average of several runs) to aggregate all abductive solutions, unless otherwise stated.

7.2 Evaluation of tabling abductive solutions

The first evaluation aims at ascertaining the relative benefit of TABDUAL's main feature, i.e. tabling abductive solutions. We employ the case of missing solutions from declarative debugging (Section 7.1.2). Consider program P_{Eval} below as buggy:

$$q_0(0,1). \qquad q_0(X,0).$$
$$q_1(1). \qquad q_1(X) \leftarrow q_0(X,X).$$
$$q_2(2). \qquad q_2(X) \leftarrow q_1(X).$$
$$q_3(3). \qquad q_3(X) \leftarrow q_2(X).$$
$$\vdots \qquad \vdots$$
$$q_{1000}(1000). \qquad q_{1000}(X) \leftarrow q_{999}(X).$$

Following the transformation for debugging missing solutions in Section 7.1.2, program P_{Eval} transforms into an abductive logic program P''_{Eval} that contains all rules of P_{Eval} plus rules below (with abducible $missing/1$):

$$q_0(X,Y) \leftarrow missing(q_0(X,Y))$$
$$q_1(X) \leftarrow missing(q_1(X))$$
$$q_2(X) \leftarrow missing(q_2(X))$$
$$\vdots$$
$$q_{1000}(X) \leftarrow missing(q_{1000}(X))$$

Program P''_{Eval} thus serves as the input for TABDUAL.

In order to evaluate tabling abductive solutions, a set of benchmarks is created that corresponds to a set of missing solutions in the buggy program P_{Eval}. More precisely, we want to debug this program for missing solutions $q_m(1001)$, where $m \in \{100, 200, \ldots, 1000\}$. Recall from Section 7.1.2, that missing a solution $q_m(1001)$, for a particular m, is expressed by adding IC $\leftarrow not\ q_m(1001)$ to the program P''_{Eval}. In our experiments, we alternatively pose query $q_m(1001)$ for each m, which is equivalent to satisfying its corresponding IC.

This set of benchmarks is suitable for showing the benefit of tabling abductive solutions. It is easy to verify that debugging missing solution $q_{100}(1001)$ obtains 101 abductive solutions: $[missing(q_{100}(1001))], [missing(q_{99}(1001))], \ldots, [missing(q_1(1001))], [missing(q_0(1001,1001))]$. By employing tabled abduction, the causes of missing solution $q_{200}(1001)$: $[missing(q_{200}(1001))], [missing(q_{199}(1001))], \ldots, [missing(q_1(1001))], [missing(q_0(1001,1001))]$, is subsequently found without recomputing those 101 abductive solutions priorly obtained from the query $q_{100}(1001)$. This advantage is accumulatively enjoyed by subsequent values of m ($m = 300, \ldots, 1000$).

Since in this first evaluation we focus on the benefit of tabling abductive solutions, we consider a variant of TABDUAL (with the same underlying implementation) where its feature of tabled abduction is stripped off. That is, by disabling the table declarations of abductive predicates $q_{i_{ab}}$, for every predicate q_i, $0 \leq i \leq 1000$. We refer below this variant as TABDUAL(NO TABLING).

TABDUAL: A TABLED ABDUCTION SYSTEM FOR LOGIC PROGRAMS

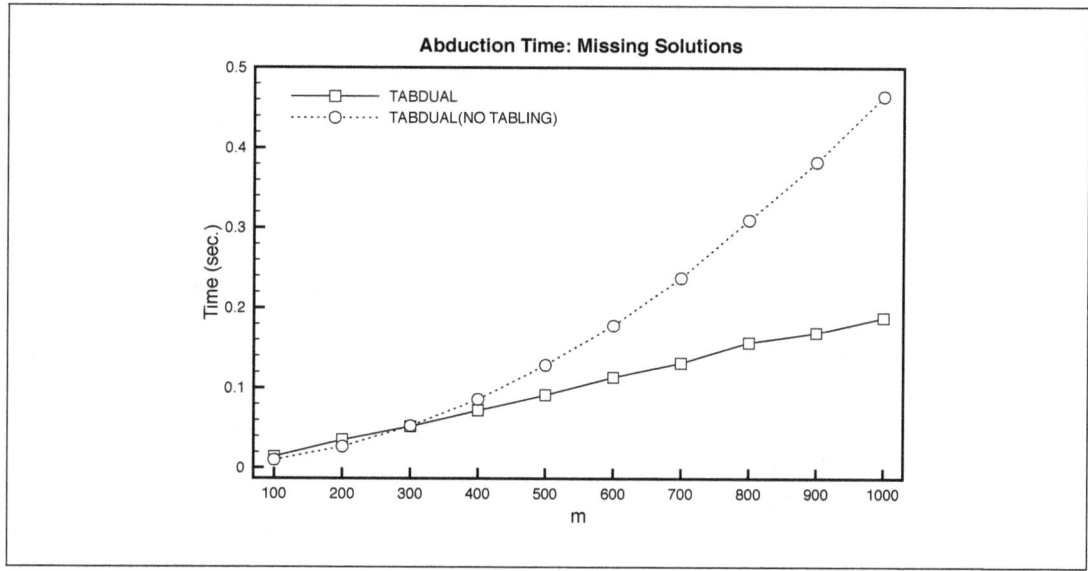

Figure 2: The abduction time for finding the causes of missing solutions $q_m(1001)$, where $m \in \{100, 200, \ldots, 1000\}$ with respect to program P_{Eval}.

Figure 2 shows the time required in the abduction stage of TABDUAL and TABDUAL(NO TABLING), by consecutively evaluating query $q_m(1001)$, where $m \in \{100, 200, \ldots, 1000\}$. Note that, the time needed in the transformation stage between the two variants do not differ, as both of them rely on the same implementation of the transformation. The result reveals that, with some little cost of tabling abductive solutions in earlier values of m (i.e. $m \leq 300$), TABDUAL consistently outperforms its counterpart in performance. Tabling pays off for subsequent values of m in TABDUAL, as greater m may reuse tabled abductive solutions of smaller m, due to the consecutive evaluation of queries. Moreover, TABDUAL scales better than TABDUAL(NO TABLING), i.e. as the values of m grows, its abduction time increases slower than the other. We also observe, that TABDUAL's abduction time tends to grow linearly, whereas that of its counterpart exponentially.

7.3 Evaluation of the dual transformations

For this evaluation, we resort to the same buggy program P_{Eval} used in the evaluation of tabling abductive solutions (cf. Section 7.2). But instead of debugging the program for missing solutions, we consider the case of incorrect solutions.

The transformation for debugging incorrect solutions (Section 7.1.1) turns P_{Eval} into an abductive logic program P'_{Eval} below (with abducible $incorrect/2$ abbreviated as $inc/2$):

$q_0(0,1) \leftarrow not\ inc(1,[0,1]).$ $q_0(X,0) \leftarrow not\ inc(2,[X,0]).$
$q_1(1) \leftarrow not\ inc(3,[1]).$ $q_1(X) \leftarrow not\ inc(4,[X]), q_0(X,X).$
$q_2(2) \leftarrow not\ inc(5,[2]).$ $q_2(X) \leftarrow not\ inc(6,[X]), q_1(X).$
$q_3(3) \leftarrow not\ inc(7,[3]).$ $q_3(X) \leftarrow not\ inc(8,[X]), q_2(X).$
$\vdots$ $\vdots$
$q_{1000}(1000) \leftarrow not\ inc(2001,[1000]).$ $q_{1000}(X) \leftarrow not\ inc(2002,[X]), q_{999}(X).$

Let us consider a set of incorrect solutions $q_m(0)$, for $m \in \{100, 200, \ldots, 1000\}$, with respect to P_{Eval}. The set of benchmarks for the purpose of this evaluation amounts to debugging each of these incorrect solutions. Recall from Section 7.1.1, that debugging an incorrect solution $q_m(0)$ for a specific m is realized by adding to program P'_{Eval} an IC $\leftarrow q_m(0)$. In our experiments, we alternatively pose query $not\ q_m(0)$ for each m, which is equivalent to satisfying its corresponding IC. Because finding the causes of incorrect solution bugs $q_m(0)$ amounts to satisfying negative goal $not\ q_m(0)$, this debugging case appropriately serves as benchmarks for evaluating dual transformations: such a negative query should be answered using dual rules computed by TABDUAL from program P'_{Eval}.

Since our aim is this particular evaluation focuses on the relative worth of the dual transformation by-need, we consider the three versions of dual transformations (Section 6.5) implemented in TABDUAL, viz.: STANDARD, BY-NEED(EAGER), and BY-NEED(LAZY). By experimenting on these three variants, we obtain results as follows:

1. It takes 1.164 seconds for TABDUAL that implements either BY-NEED(EAGER) or BY-NEED(LAZY), whereas the implementation STANDARD 1.674 seconds. It is obvious that BY-NEED(EAGER) and BY-NEED(LAZY) require less transformation time than STANDARD, since they do not produce all dual rules in advance as STANDARD does.

   Take an example q_2. The second layer dual rules produced, in advance, by STANDARD are: (apart from the dual rule that disunifies arguments $q_2^{*1}(X,I,I) \leftarrow X \neq 2$):

   $q_2^{*1}(2,I,O) \leftarrow incorrect(5,[2],I,O).$
   $q_2^{*2}(X,I,O) \leftarrow incorrect(6,[X],I,O).$
   $q_2^{*2}(X,I,O) \leftarrow not\_q_1(X,I,O).$

   whereas BY-NEED(EAGER) and BY-NEED(LAZY) just produce their skeleton:

   $q_2^{*1}(2,I,O) \leftarrow dual(1,q_2(2),I,O).$
   $q_2^{*2}(X,I,O) \leftarrow dual(2,q_2(X),I,O).$

   and only construct dual rules, by-need, during abduction.

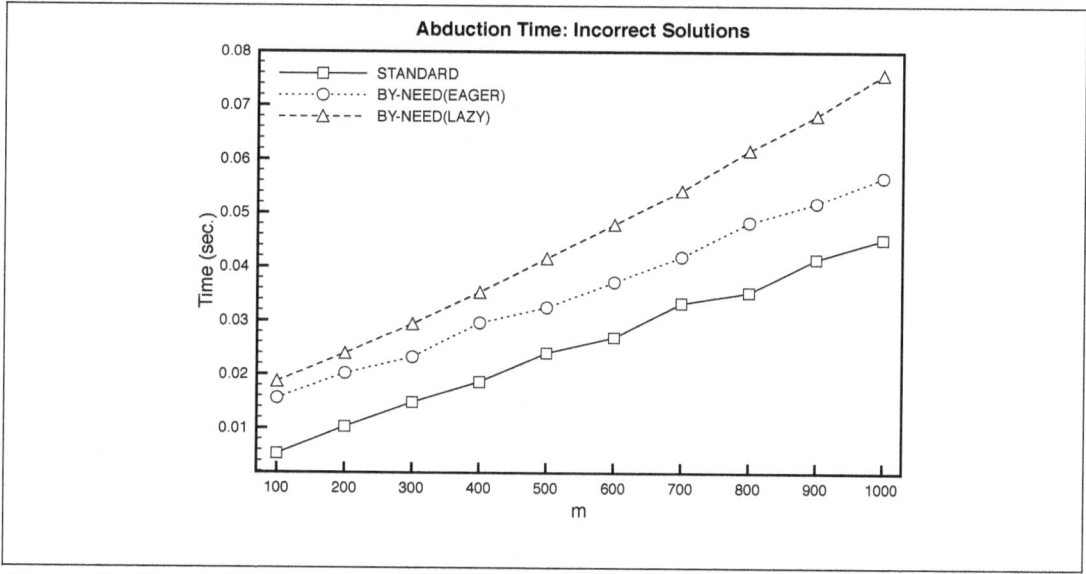

Figure 3: The abduction time for finding the causes of incorrect solutions $q_m(0)$, where $m \in \{100, 200, \ldots, 1000\}$ with respect to program P_{Eval}.

2. In terms of the number of dual rules (apart from those dual rules defined by disunifying arguments), STANDARD creates 3002 second layer dual rules during the transformation, regardless their need. On the other hand, BY-NEED(EAGER) and BY-NEED(LAZY) create only 2002 (skeleton of) second layer dual rules, shown above.

3. During abduction, BY-NEED(EAGER) and BY-NEED(LAZY) construct only 60% of second layer dual rules produced by STANDARD. That is, 40% of dual rules constructed by STANDARD are actually not needed. Take again the above case of q_2. In finding the causes of an incorrect solution $q_m(0)$ (for some m), query $not\ q_m(0)$ does not need to invoke dual rule $q_2^{*1}(2, I, O) \leftarrow incorrect(5, [2], I, O)$, as they are bound to fail. In this case, the other dual rule that disunifies arguments, viz., $q_2^{*1}(X, I, I) \leftarrow X \neq 2$, already succeeds (as X is instantiated by 0) and is sufficient to satisfy the q_2^{*1} part.

Point (1) above tells us that the dual transformation by-need, either BY-NEED(EAGER) or BY-NEED(LAZY), requires less time in the transformation stage compared to STANDARD. On the other hand, as explained in Section 6.5 (cf. explanation for Example 17), predicate $dual/4$, in the rule skeleton of point (2) above, constructs generic dual rules and instantiates them only in the abduction stage. Figure 3 shows the performance of each dual transformation in terms of time needed in the abduction stage of consecutively running query $not\ q_m(0)$ with different values of $m \in \{100, 200, \ldots, 1000\}$. Note that, the time needed for the transformation is not included in the figure, as we want to measure the cost incurred by BY-NEED(EAGER) and BY-NEED(LAZY) in the abduction stage. In Figure 3, we observe that STANDARD is faster than BY-NEED(EAGER) and BY-NEED(LAZY). This is expected, due to the overhead incurred for computing dual rules on-the-fly, by need, during the abduction stage of BY-NEED(EAGER) and BY-NEED(LAZY).

On the other hand, if we consider the whole process (i.e., the time needed in the transformation plus the abduction stages), the time overhead of BY-NEED(EAGER) and BY-NEED(LAZY) in the abduction stage is well-compensated by their transformation time. That is, the time for the whole process (transformation plus abduction) of STANDARD is 1.929 seconds, whereas BY-NEED(EAGER) and BY-NEED(LAZY) need 1.521 and 1.620 seconds, respectively.

In this evaluation scenario, where all abductive solutions are aggregated, the performance of BY-NEED(LAZY) is slightly worse than BY-NEED(EAGER). This can be explained by the extra maintenance of the tracking information needed for the explicit memoization in BY-NEED(LAZY). It may as well explain that the time gap between BY-NEED(LAZY) and BY-NEED(EAGER) is wider as m grows, meaning more dual rules are stored in the trie. Nevertheless, BY-NEED(LAZY) returns the first abductive solution much faster than BY-NEED(EAGER), e.g. at $m = 1000$ the lazy one needs 0.0003 seconds, whereas the eager one 0.0146 seconds. Aggregating all solutions may not be a realistic scenario in abduction as one cannot wait indefinitely for all solutions, whose number might even be infinite. Instead, one chooses a solution that satisfices so far, and may continue searching for more, if needed. In that case, it seems reasonable that the lazy dual rules computation may be competitive against the eager one. Nevertheless, the two approaches are available as options for TABDUAL customization.

7.4 Evaluation of tabling nogoods of subproblems

The technique of recording nogoods of subproblems, i.e. inconsistent solutions of subproblems that cannot be extended to derive any solution of the given problem, has been employed in truth maintenance systems [10, 13], constraint satisfaction problems [45], SAT solvers [27], and in answer set solvers [19], to help prune search space.

We employ TABDUAL and show that tabling abductive solutions can be appropriate for tabling nogoods of subproblems. For this purpose, we consider the well-known N-queens problem, where abduction is used to find safe board configurations of N queens. The problem is represented in TABDUAL as follows:

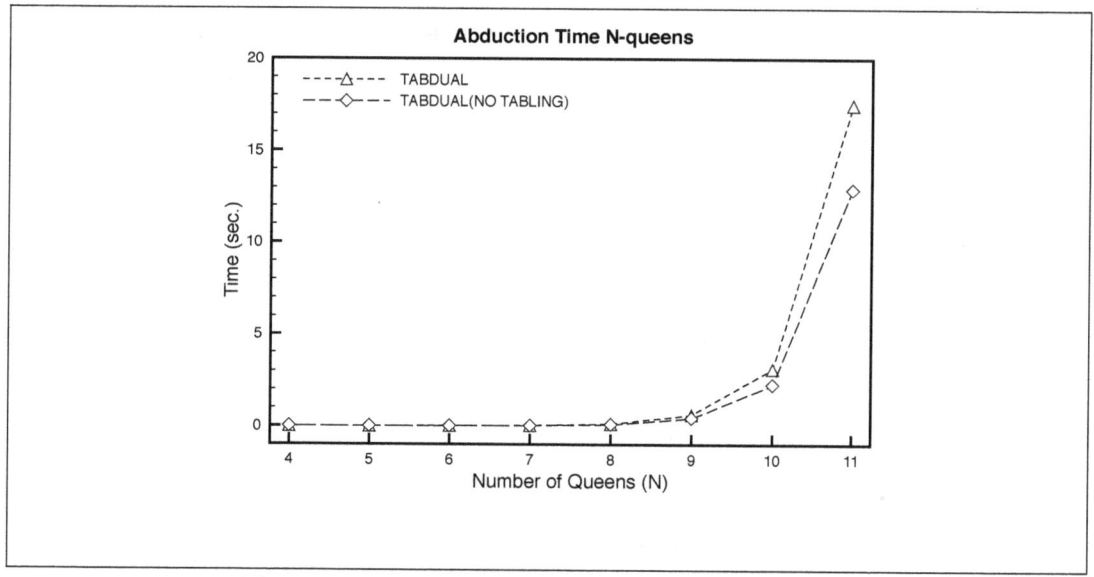

Figure 4: The abduction time of different N queens.

$$q(0, N).$$
$$q(M, N) \leftarrow M > 0,\ q(M-1, N),\ d(Y),\ pos(M, Y),$$
$$process\_ongoing(not\ conflict).$$

$$conflict(Conf) \leftarrow prolog(conflicting(Conf)).$$

and the query is $q(N, N)$ for N queens. Here, $pos/2$ is the abducible representing the position of a queen, and $d/1$ is a column generator predicate, available as facts $d(i)$ for $1 \leq i \leq N$. Predicate $conflicting/1$ is defined in a non-abductive program module, to check whether the ongoing board configuration $Conf$ of queens is conflicting. By scaling up the problem, i.e. increasing the value of N, we aim at evaluating the scalability of TABDUAL, concentrating on tabling nogoods of subproblems (essentially, tabling nogoods for use by ongoing abductive solutions); in this case, it means tabling conflicting configurations of queens.

Since this benchmark is used to evaluate the benefit of *tabling* nogoods of subproblems (as abductive solutions), and *not* the benefit of the dual by-need improvement, we evaluate TABDUAL compared to its non-tabling variant TABDUAL(NO TABLING), as in Section 7.2. The transformation time of the problem representation is similar for both of them, i.e. around 0.003 seconds. Figure 4 shows abduction time for N queens, $4 \leq N \leq 11$. The reason that TABDUAL performs worse than TABDUAL(NO TABLING) is that the con-

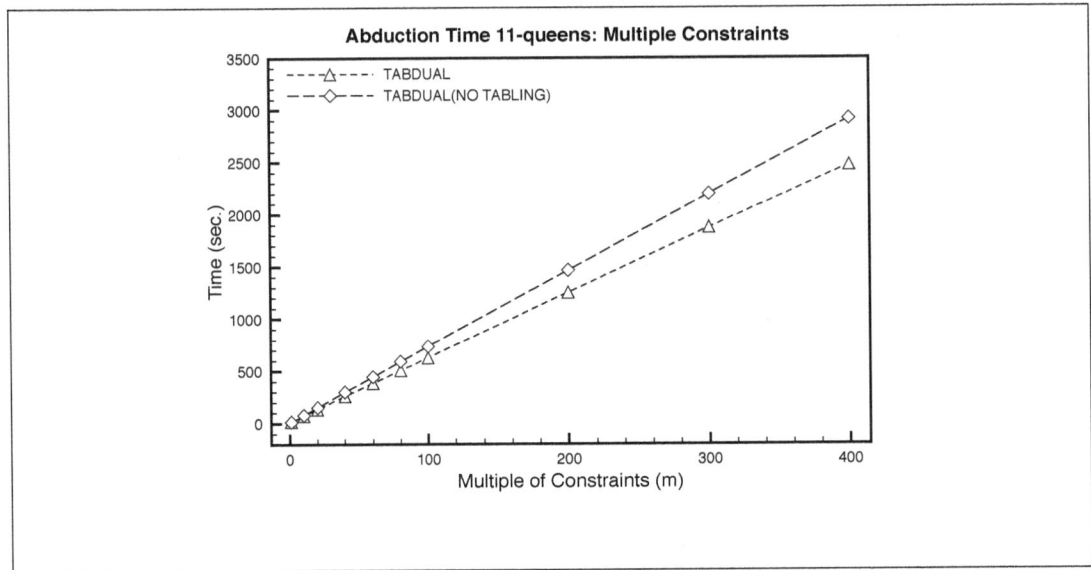

Figure 5: The abduction time of 11 queens with increasing complexity of conflict constraints.

flict constraints in the N-queens problem are quite simple, i.e. consist of only column and diagonal checking. It turns out that tabling such simple conflicts does not pay off, that the cost of tabling overreaches the cost of Prolog recomputation. But what if we increase the complexity of the constraints, e.g. adding more queen's attributes (colors, shapes, etc.) to further constrain its safe positioning?

Figure 5 shows abduction time for 11 queens with increasing complexity of the conflict constraints. To simulate different complexity, the conflict constraints are repeated m number of times, where m varies from 1 to 400. It shows that TABDUAL's performance is remedied and, benefitting from tabling the ongoing conflict configurations, it consistently surpasses the performance of TABDUAL(NO TABLING), showing increasing improvement as m increases, up to 15% for $m = 400$. That is, it is scale consistent with respect to the complexity of the constraints.

7.5 Evaluation of programs with loops

We also evaluate TABDUAL for programs with loops in the presence of tabled abduction, as detailed in Section 6.4. For that purpose, we employ a set of ground programs with various combination of loops, many of which cover difficult known cases of such programs. The test-suite has previously been used in evaluating the ABDUAL meta-interpreter in [5].

Consider the following six ground programs from the test-suite:

$p_0 \leftarrow q_0.$ $\quad p_3 \leftarrow q_3.$ $\quad p_4 \leftarrow q_4.$
$p_0 \leftarrow a.$ $\quad q_3 \leftarrow not\ r_3.$ $\quad q_4 \leftarrow p_4.$
$q_0 \leftarrow p_0.$ $\quad r_3 \leftarrow p_3.$ $\quad q_4 \leftarrow not\ a,\ not\ b.$
$q_0 \leftarrow b.$

$p_8 \leftarrow not\ q_8,\ a.$ $\quad p_{11} \leftarrow not\ q_{11},\ a.$
$q_8 \leftarrow not\ p_8.$ $\quad q_{11} \leftarrow p_{11},\ not\ a.$
$q_8 \leftarrow b.$

where a and b are abducibles.

We provide a comparison of the results returned by both implementations, focusing particularly on those that differ. Table 1 lists the answers of the given queries to the corresponding programs, returned by TABDUAL and the ABDUAL meta-interpreter. It shows that TABDUAL returns correct answers according to the ABDUAL theory which underpins it, even now in the presence of tabled abduction. On the other hand, in the cases we are showing here, the ABDUAL meta-interpreter does not correctly return answers as computed by its theory. The answers returned by TABDUAL for queries in Table 1 are explained as follows:

- For query $not\ p_0$, $[not\ a, not\ b]$ should be the only solution, because $not\ p_0$ succeeds by abducing $not\ a$ *and* failing q_0. To fail q_0, $not\ b$ has to be abduced *and* p_0 has to fail. Here, there is a positive loop on negation between $not\ p_0$ and $not\ q_0$, so the query succeeds and gives the solution $[not\ a, not\ b]$ as the only solution.

- For queries p_3 and $not\ p_3$, unlike answers returned by the ABDUAL meta-interpreter, TABDUAL returns *undefined* (and abduces nothing) as expected, due to the negative loops over negation.

- Query $not\ p_4$ shows that TABDUAL does less abduction than the ABDUAL meta-interpreter, by abducing a or b only, but not both.

- For query q_8, TABDUAL has an additional solution [ ], i.e. nothing is abduced, making particularly a false and consequently p_8 false (or, $not\ p_8$ true). Thus, query q_8 is true (by its first rule) under this solution, which is missing in answers returned by the ABDUAL meta-interpreter.

- For query $not\ q_{11}$, the first solution is obtained by abducing a to fail q_{11}. Another way to fail q_{11} is to fail p_{11}, which gives another solution, by abducing $not\ a$. These are the only two abductive solutions which are returned by TABDUAL and follows correctly

Queries	TABDUAL	ABDUAL meta-interpreter
$not\ p_0$	$[not\ a, not\ b]$	$[not\ a, not\ b], [not\ a]$
p_3	$[\ ]$ undefined	$[\ ]$
$not\ p_3$	$[\ ]$ undefined	$[\ ]$
$not\ p_4$	$[a], [b]$	$[a], [b], [a, b]$
q_8	$[\ ], [not\ a], [b]$	$[not\ a], [b]$
$not\ q_{11}$	$[a], [not\ a]$	$[\ ], [a], [not\ a]$

Table 1: Answers by TABDUAL vs. ABDUAL meta-interpreter.

the definition of abductive solutions. There is no direct positive loop involving q_{11} in the program, hence $not\ q_{11}$ will never succeed with the $[\ ]$ abductive solution, as returned by the ABDUAL meta-interpreter.

In addition to ground programs, we also evaluate TABDUAL on non-ground programs, i.e. programs having variables (with or without loops), which is not afforded by ABDUAL. The latter system does not allow rules having variables, i.e. rules with variables in a program have first to be ground with respect to the Herbrand universe (like in answer set programming systems). The complete test-suite and the evaluations results are detailed in Appendix B.

8 Concluding remarks

We have addressed the issue of tabling abductive solutions, in a way that they can be reused from one abductive context to another. We do so by resorting to a program transformation approach, resulting in a tabled abduction prototype, TABDUAL, implemented in XSB Prolog. TABDUAL is underpinned by the ABDUAL theory and employs its dual transformation, which allows to more efficiently handle the problem of abduction under negative goals. In TABDUAL, abducibles are treated much like terminals in grammars, with an extra argument for input and another for output abductive context accumulation. A few other original innovative and pragmatic techniques are employed to handle programs with variables and loops, as well as to make TABDUAL more efficient and flexible. It has been evaluated with various objectives in mind, in order to show the benefit of tabled abduction and to gauge its suitability for likely applications. An issue that we have touched upon in the TABDUAL evaluation is that of tabling nogoods of subproblems in the context of tabled abduction, and how it may improve performance and scalability. The other evaluation result reveals that each approach of the dual transformation by-need may be suitable for different situations, i.e. both approaches, BY-NEED(EAGER) or BY-NEED(LAZY), are options for TABDUAL customization.

8.1 Related work

There have been a plethora of work on abduction in logic programming, cf. [12, 22] for a survey on this line of work. But, with the exception of ABDUAL [4], we are not aware of any other efforts that have addressed the use of tabling in abduction for abductive normal logic programs, which may be complicated with loops. Like ABDUAL, we use the dual transformation and rely on the same theoretic underpinnings, but ABDUAL does not allow variables in rules. The reader is referred to Section 5.2 of [4] on how the dual transformation and its properties relate to other works.

From the implementation viewpoint, tabling has only been employed in ABDUAL limitedly, i.e. to table its meta-interpreter, which in turn allows abduction to be performed (also in the presence of loops in a program), but it does not address at all the specific issues raised by the desirable reuse of tabled abductive solutions. TABDUAL generates a self-sufficient program transform (plus system predicates), which employs no meta-interpreter, even in the presence of loops in programs.

Our approach also differs from that of [2]. Therein, abducibles are coded as odd loops, it is compatible with and uses constructive negation, and it involves manipulating the residual program. It suffers from a number of problems, which it identifies, in its Sections 5 and 6, and its approach was not pursued further.

TABDUAL does not concern itself with constructive negation, like NEGABDUAL [3] and its follow-up [7]. NEGABDUAL uses abduction to provide constructive negation plus abduction, by making the disunification predicate an abducible. Again, it does not concern itself with the issues of tabled abductive solution reuse, which is the main purpose of TABDUAL. However, because of its constructive negation ability, NEGABDUAL can deal with problems that TABDUAL does not. Consider program below, with no abducibles, just to illustrate the point of constructive negation induced by dualization:

$$p(X) \leftarrow q(Y). \qquad q(1).$$

In NEGABDUAL, the query $not\ p(X)$ will return a qualified '*yes*', because it is always possible to solve the constraint $Y \neq 1$, as long as one assumes there are at least two constants in the Herbrand Universe. Distinct from NEGABDUAL, TABDUAL answers '*no*' to $not\ p(X)$. It is correct, in the absence of conditional answers; the former answer is afforded only by having constructive negation in place. It is interesting to explore in the future, whether TABDUAL can be extended to take care such constraints (as abducibles), given that XSB supports low-level constraint handling through attributed variables, and that attributed variables can be tabled in XSB.

TABDUAL, being implemented in XSB, is based on the WFS, which enjoys the relevance property, induced by the top-down query-oriented procedure, solely for finding the relevant abducibles and their truth value. This is not the case with the bottom-up approaches for abduction, e.g. [43], where stable models for computing abductive explanations, not necessarily related to an observation, are constructed. This disadvantage of the bottom-up TMS approach is in fact later avoided by adding a top-down procedure, as in [44]. TABDUAL also allows dealing with odd loops in programs because of its 3-valued program semantics, whilst retaining 2-valued abduction and the use of integrity constraints. This is not enjoyed by the bottom-up approach and its 2-valued implementation.

The tabling technique, within the context of statistical abduction, is employed in [42]. But it concerns itself with probabilistic logic programs, whereas TABDUAL concerns abductive normal logic programs. Moreover, the tabling technique in [42] imposes the so-called 'acyclic support condition', a constraint that does not allow loops in a program, which pose no restrictions at all in TABDUAL. Tabling is also used recently in PITA [35], for statistical abduction. Though PITA is also based on the Well-Founded Semantics like TABDUAL, tabling (in particular its feature, answer subsumption) applies specifically to probabilistic logic programs, e.g. to compute the number of different explanations for a subgoal (in terms of Viterbi path), which is not our concern in TABDUAL, and thus does not employ the dual transformation and other techniques described here.

8.2 Future work

TABDUAL still has much room for improvement. Future work will consist in continued exploration of our applications of abduction, which will provide feedback for system improvement. Tabled abduction may benefit from answer subsumption [47] in tabling abductive solutions to deal with redundant explanations, in the sense that it suffices to table only smaller abductive solutions (with respect to the subset relation). Another potential XSB feature to look into is the applicability of interning ground terms [52] for tabling abductive solutions, which are ground, and study how extra efficiency may be gained from it.

The implementation technique of BY-NEED(LAZY) consists in operational details that are facilitated by XSB's trie manipulation predicates, to simulate the batched-like table scheduling within XSB's current default local table scheduling. In order to have a more transparent implementation of those operations, it is desirable that XSB permits a mixture in using batched and local table scheduling strategies, or alternatively, stopping the evaluation at some first answers to a subgoal within the currently default local table scheduling.

In another related research line, it would be interesting to explore whether abduction could be used with XSB's partial support of Transaction Logic in its *storage* module. There is a certain similarity in that Transaction Logic rules may cause updates in a manner that is reminiscent of abduction, although Transaction Logic also allows a commit.

TABDUAL opens up a potential joint use with other non-monotonic LP features, having their own tabling requirements and attending benefits. In [40], we combine tabled abduction with LP updating [37]. The latter employs incremental tabling to automatically propagate updates bottom-up, being triggered by a top-down query. The implemented system of this joint LP abduction and updating is part and parcel of our research to employ LP in agent moral reasoning [41]. We envisage a couple of applications using this joint system, and tabled abduction particularly, in this field. For one, people often engage counterfactual thoughts in moral situations, where their function is not just evaluative (to correct wrong behavior in the past), but also reflective (to simulate possible alternatives for a careful consideration before making a moral decision) [15]. In [33], we propose a LP approach to model counterfactual reasoning, based on Pearl's structural approach [28] (abstaining from probability), by benefitting from the aforementioned joint system. The role of abduction is this LP counterfactual approach is to hypothesize background conditions from given evidences or observations, a required step in Pearl's approach, so as to provide an initial abductive context for the counterfactual being evaluated. In this case, tabled abduction permits reusing this initial abductive context in another subsequent context acquired by the agent during its life cycle.

Another morality related application we consider is the dual-process model in moral judgment [8] that stresses the interaction between deliberative and reactive behaviors in delivering moral judgment. Given that abductive solutions represent some actions according to a specific moral principle, tabled abduction may play several roles. For one thing, it allows an agent to deliver an action in exactly the same context without repeating the same deliberative reasoning, thus simulating a form of low-level reactive behavior (realized by system-level tabling) of the dual-process model. For another, in a dynamic environment (hence the need of LP updating), the agent may later be required to achieve new goals in addition to the former ones, due to a moral principle it follows. While achieving these new goals requires deliberative reasoning, the decisions that have been abduced for former goals can immediately be retrieved from the table and are subsequently involved in (as the context of) the deliberative reasoning for the new goals. It thus provides a computational model of collaborative interaction between deliberative and reactive reasoning in the dual-process model.

Abduction is by now a staple feature of hypothetical reasoning and non-monotonic knowledge representation. It is already mature enough in its concept, deployment, applications, and proof-of-principle, to warrant becoming a run-of-the-mill ingredient in a Logic Programming environment. We hope this work will lead, in particular, to an XSB System that can provide its users with specifically tailored tabled abduction facilities, by migrating some of TABDUAL's features to its engine level.

References

[1] M. Alberti, F. Chesani, M. Gavanelli, E. Lamma, and P. Torroni. Security protocols verification in abductive logic programming. In *6th Int. Workshop on Engineering Societies in the Agents World (ESAW)*, volume 3963 of *LNCS*. Springer, 2005.

[2] J. J. Alferes and L. M. Pereira. Tabling abduction. 1st Intl. Ws. Tabulation in Parsing and Deduction (TAPD'98), http://centria.di.fct.unl.pt/~lmp/publications/online-papers/tapd98abd.ps.gz, 1998.

[3] J. J. Alferes and L. M. Pereira. NEGABDUAL meta-interpreter. Available from http://centria.di.fct.unl.pt/~lmp/software/contrNeg.rar, 2007.

[4] J. J. Alferes, L. M. Pereira, and T. Swift. Abduction in well-founded semantics and generalized stable models via tabled dual programs. *Theory and Practice of Logic Programming*, 4(4):383–428, 2004.

[5] J. J. Alferes, L. M. Pereira, and T. Swift. ABDUAL meta-interpreter. Available from http://www.cs.sunysb.edu/~tswift/interpreters.html, 2004.

[6] J. Balsa, V. Dahl, and J. G. Pereira Lopes. Datalog grammars for abductive syntactic error diagnosis and repair. In *Proc. Natural Language Understanding and Logic Programming Workshop*, 1995.

[7] V. P. Ceruelo. Negative non-ground queries in well founded semantics. Master's thesis, Universidade Nova de Lisboa, 2009.

[8] F. Cushman, L. Young, and J. D. Greene. Multi-system moral psychology. In J. M. Doris, editor, *The Moral Psychology Handbook*. Oxford University Press, 2010.

[9] C. V. Damásio and L. M. Pereira. Abduction over 3-valued extended logic programs. In *Procs. 3rd. Intl. Conf. Logic Programming and Non-Monotonic Reasoning (LPNMR)*, volume 928 of *LNAI*, pages 29–42. Springer, 1995.

[10] J. de Kleer. An assumption-based TMS. *Artificial Intelligence*, 28(2):127–162, 1986.

[11] M. Denecker and D. de Schreye. SLDNFA: An abductive procedure for normal abductive programs. In *Procs. of the Joint Intl. Conf. and Symp. on Logic Programming*. The MIT Press, 1992.

[12] M. Denecker and A. C. Kakas. Abduction in logic programming. In *Computational Logic: Logic Programming and Beyond*. Springer Verlag, 2002.

[13] J. Doyle. A truth maintenance system. *Artificial Intelligence*, 12:231–272, 1979.

[14] T. Eiter, G. Gottlob, and N. Leone. Abduction from logic programs: semantics and complexity. *Theoretical Computer Science*, 189(1-2):129–177, 1997.

[15] K. Epstude and N. J. Roese. The functional theory of counterfactual thinking. *Personality and Social Psychology Review*, 12(2):168–192, 2008.

[16] K. Eshghi. Abductive planning with event calculus. In *Proc. Intl. Conf. on Logic Programming*. The MIT Press, 1988.

[17] T. H. Fung and R. Kowalski. The IFF procedure for abductive logic programming. *Journal of Logic Programming*, 33(2):151–165, 1997.

[18] J. Gartner, T. Swift, A. Tien, C. V. Damásio, and L. M. Pereira. Psychiatric diagnosis from the

viewpoint of computational logic. In *Procs. 1st Intl. Conf. on Computational Logic (CL 2000)*, volume 1861 of *LNAI*, pages 1362–1376. Springer, 2000.

[19] M. Gebser, B. Kaufmann, A. Neumann, and T. Schaub. Conflict-driven answer set solving. In *Procs. 20th Intl. Joint Conf. on Artificial Intelligence (IJCAI)*, 2007.

[20] K. Inoue and C. Sakama. A fixpoint characterization of abductive logic programs. *J. of Logic Programming*, 27(2):107–136, 1996.

[21] J. R. Josephson and S. G. Josephson. *Abductive Inference: Computation, Philosophy, Technology*. Cambridge U. P., 1995.

[22] A. Kakas, R. Kowalski, and F. Toni. The role of abduction in logic programming. In D. Gabbay, C. Hogger, and J. Robinson, editors, *Handbook of Logic in Artificial Intelligence and Logic Programming*, volume 5. Oxford U. P., 1998.

[23] A. C. Kakas and P. Mancarella. Knowledge assimilation and abduction. In *Intl. Workshop on Truth Maintenance*, ECAI'90, 1990.

[24] A. C. Kakas and A. Michael. An abductive-based scheduler for air-crew assignment. *J. of Applied Artificial Intelligence*, 15(1-3):333–360, 2001.

[25] R. Kowalski and F. Sadri. Abductive logic programming agents with destructive databases. *Annals of Mathematics and Artificial Intelligence*, 62(1):129–158, 2011.

[26] P. Lipton. *Inference to the Best Explanation*. Routledge, 2001.

[27] J. P. Marques-Silva and K. A. Sakallah. GRASP: A new search algorithm for satisfiability. In *International Conference on Computer-Aided Design*, pages 220–227, 1996.

[28] J. Pearl. *Causality: Models, Reasoning and Inference*. Cambridge U. P., 2009.

[29] L. M. Pereira, C. V. Damásio, and J. J. Alferes. Debugging by diagnosing assumptions. In *Automatic Algorithmic Debugging*, volume 749 of *LNCS*, pages 58–74. Springer, 1993.

[30] L. M. Pereira, C. V. Damásio, and J. J. Alferes. Diagnosis and debugging as contradiction removal in logic programs. In *Progress in Artificial Intelligence*, volume 727 of *LNAI*, pages 183–197. Springer, 1993.

[31] L. M. Pereira, P. Dell'Acqua, A. M. Pinto, and G. Lopes. Inspecting and preferring abductive models. In K. Nakamatsu and L. C. Jain, editors, *The Handbook on Reasoning-Based Intelligent Systems*, pages 243–274. World Scientific Publishers, 2013.

[32] L. M. Pereira and A. Saptawijaya. Abductive logic programming with tabled abduction. In *ICSEA 2012*, pages 548–556. ThinkMind, 2012.

[33] L. M. Pereira and A. Saptawijaya. Counterfactuals in logic programming with applications to agent morality. Available from http://centria.di.fct.unl.pt/~lmp/publications/online-papers/moral_counterfactuals.pdf, 2015.

[34] O. Ray, A. Antoniades, A. Kakas, and I. Demetriades. Abductive logic programming in the clinical management of HIV/AIDS. In *Proc. 17th. European Conference on Artificial Intelligence*. IOS Press, 2006.

[35] F. Riguzzi and T. Swift. The PITA system: Tabling and answer subsumption for reasoning under uncertainty. *Theory and Practice of Logic Programming*, 11(4-5):433–449, 2011.

[36] A. Saptawijaya and L. M. Pereira. Implementing tabled abduction in logic programs. In *Procs. 16th Portuguese Intl. Conf. on Artificial Intelligence (EPIA)*, Doctoral Symposium on Artificial

Intelligence (SDIA), 2013.

[37] A. Saptawijaya and L. M. Pereira. Incremental tabling for query-driven propagation of logic program updates. In *LPAR-19*, volume 8312 of *LNCS*, pages 694–709. Springer, 2013.

[38] A. Saptawijaya and L. M. Pereira. Towards practical tabled abduction in logic programs. In *16th Portuguese Conference on Artificial Intelligence (EPIA)*, LNAI. Springer, 2013.

[39] A. Saptawijaya and L. M. Pereira. Towards practical tabled abduction usable in decision making. In *KES-IDT 2013*, Frontiers of Artificial Intelligence and Applications (FAIA). IOS Press, 2013.

[40] A. Saptawijaya and L. M. Pereira. Joint tabling of logic program abductions and updates (Technical Communication of ICLP 2014). *Theory and Practice of Logic Programming, Online Supplement*, 14(4-5), 2014. Available from http://arxiv.org/abs/1405.2058.

[41] A. Saptawijaya and L. M. Pereira. The potential of logic programming as a computational tool to model morality. In R. Trappl, editor, *A Construction Manual for Robots' Ethical Systems: Requirements, Methods, Implementations*, Cognitive Technologies (forthcoming). Springer, 2015. http://centria.di.fct.unl.pt/~lmp/publications/online-papers/ofai_book.pdf.

[42] T. Sato and Y. Kameya. Parameter learning of logic programs for symbolic-statistical modeling. *J. of Artificial Intelligence Research (JAIR)*, 15:391–454, 2001.

[43] K. Satoh and N. Iwayama. Computing abduction by using the TMS. In *Procs. 8th Intl. Conf. on Logic Programming (ICLP)*, pages 505–518. The MIT Press, 1991.

[44] K. Satoh and N. Iwayama. Computing abduction by using TMS and top-down expectation. *Journal of Logic Programming*, 44(1-3):179–206, 2000.

[45] T. Schiex and G. Verfaillie. Nogood recording for static and dynamic constraint satisfaction problems. In *Procs. 5th. Intl. Conf. on Tools with Artificial Intelligence (ICTAI)*, 1993.

[46] T. Swift. Tabling for non-monotonic programming. *Annals of Mathematics and Artificial Intelligence*, 25(3-4):201–240, 1999.

[47] T. Swift and D. S. Warren. Tabling with answer subsumption: Implementation, applications and performance. In *JELIA 2010*, volume 6341 of *LNCS*, pages 300–312. Springer, 2010.

[48] T. Swift and D. S. Warren. XSB: Extending Prolog with tabled logic programming. *Theory and Practice of Logic Programming*, 12(1-2):157–187, 2012.

[49] T. Swift and D. S. Warren. Personal communications, February 2012, 2013.

[50] T. Swift, D. S. Warren, K. Sagonas, J. Freire, P. Rao, B. Cui, E. Johnson, L. de Castro, R. F. Marques, D. Saha, S. Dawson, and M. Kifer. *The XSB System Version 3.3.x Volume 1: Programmer's Manual*, 2012.

[51] A. van Gelder, K. A. Ross, and J. S. Schlipf. The well-founded semantics for general logic programs. *Journal of ACM*, 38(3):620–650, 1991.

[52] D. Warren. Interning ground terms in XSB. In *Colloquium on Implementation of Constraint and Logic Programming Systems (CICLOPS 2013)*, 2013.

A Proof of Theorem 2

Theorem 1. *Let P be an abductive normal logic program and $\mathcal{A}_P$ be the set of abducible atoms in P. Then $size(\tau(P)) < 9.size(P) + 4.|\mathcal{A}_P|$.*

Proof. Let p_i be a predicate for which there are $m > 0$ rules in P with the total size $size(P|_{p_i})$, and $c \geq 0$ be the number of abducibles in the body of a rule of p_i.

- Since the abducibles in the body of a rule are moved from the body to abductive context (cf. point (1) of Definition 1), we have the size of $\tau'(P)$ as $size(\tau'(P)) = size(P) - c.rules(P)$.

- Since $\tau^+(p_i)$ for every defined $p_i \in P$ has three literals (cf. point (2) of Definition 1), we have the size of $\tau^+(P)$ as $size(\tau^+(P)) = 3.heads(P)$.

- For τ^-, we have two cases, based on Definition 2:
  1. By point 1(a), i.e. for p_i defined in P, the size of $\tau^-(p_i)$ will be $m + 1$. Therefore, the size of the transformed program in P by τ^- for all predicates defined in P will be $heads(P).(m+1) = rules(P) + heads(P)$.
  2. By point 2, the size of the transformed program in P by τ^- for all predicates that have no definition in P will be $preds(P) - heads(P)$.

  Summing up the size from both cases, we have:
  $size(\tau^-(P)) = rules(P) + preds(P)$.

- For τ^*, the total size of rules with heads of the form $p^{*i}(\bar{t}_i, I, O)$, cf. point 1(b) of Definition 2, will be $2.(size(P|_{p_i}) - m)$. Since the size of the other rule, i.e. the one with the head $p^{*i}(\bar{X}, I, I)$, is two (for each i), the total size of $\tau^*(p_i)$ is $2.(size(P|_{p_i}) - m) + 2m = 2.size(P|_{p_i})$. Therefore, the size of the transformed program in P by τ^* for all predicates defined in P will be $size(\tau^*(P)) = \Sigma_{i=1}^{heads(P)} 2.size(P|_{p_i}) = 2.size(P)$.

- Finally for τ°, since the size of rules for each abducible atom is four, we have $size(\tau^\circ(P)) = 4.|\mathcal{A}_P|$, where $|\mathcal{A}_P|$ denotes the cardinality of $\mathcal{A}_P$.

Note that $preds(P) \leq size(P)$, as also for $heads(P)$ and $rules(P)$. Thus:
$size(\tau(P)) = size(\tau'(P)) + size(\tau^+(P)) + size(\tau^-(P)) + size(\tau^*(P)) + size(\tau^\circ(P)) < 9.size(P) + 4.|\mathcal{A}_P|$. □

$$p_0 \leftarrow q_0.$$
$$p_0 \leftarrow a.$$
$$q_0 \leftarrow p_0.$$
$$q_0 \leftarrow b.$$

$$p_1 \leftarrow not\ q_1, r_1.$$
$$r_1 \leftarrow not\ q_1, p_1.$$
$$q_1 \leftarrow not\ p_1.$$

$$p_2 \leftarrow q_2.$$
$$q_2 \leftarrow r_2.$$
$$r_2 \leftarrow p_2.$$

$$p_3 \leftarrow q_3.$$
$$q_3 \leftarrow not\ r_3.$$
$$r_3 \leftarrow p_3.$$

$$p_4 \leftarrow q_4.$$
$$q_4 \leftarrow p_4.$$
$$q_4 \leftarrow not\ a, not\ b.$$

$$p_5 \leftarrow q_5.$$
$$q_5 \leftarrow not\ r_5.$$
$$r_5 \leftarrow not\ s_5.$$
$$s_5 \leftarrow p_5.$$

$$p_6 \leftarrow not\ q_6.$$
$$q_6 \leftarrow r_6.$$
$$r_6 \leftarrow s_6.$$
$$s_6 \leftarrow not\ p_6.$$

$$p_7 \leftarrow not\ q_7, r_7, a.$$
$$r_7 \leftarrow not\ q_7, p_7, b.$$
$$q_7 \leftarrow not\ p_7, not\ r_7.$$

$$p_8 \leftarrow not\ q_8, a.$$
$$q_8 \leftarrow not\ p_8.$$
$$q_8 \leftarrow b.$$

$$p_{10} \leftarrow not\ q_{10}, a.$$
$$q_{10} \leftarrow p_{10}, a.$$

$$p_{11} \leftarrow not\ q_{11}, a.$$
$$q_{11} \leftarrow p_{11}, not\ a.$$

$$p_{12} \leftarrow a, not\ q_{12}.$$
$$q_{12} \leftarrow not\ a, p_{12}.$$

Figure 6: Collection of Ground Programs with Loops

B Test-suite

The test-suite consists of two collections of programs: ground programs with loops, and programs with variables (also containing loops). In the evaluation of ground programs with loops, a comparison with the ABDUAL meta-interpreter [5] is made. Both systems run on the same platform under XSB version 3.3.7.

B.1 Programs with loops

Collection of programs Figure 6 lists a collection of programs, including difficult cases, used to compare TABDUAL and the ABDUAL meta-interpreter. The collection is specific to ground programs, since ABDUAL caters only to ground programs and queries. The evaluation results are shown subsequently. These programs involve various loops: direct positive loops, negative loops over negation, positive loops in (dualized) negation, and some combinations amongst them. In this collection, a, b, and c are abducibles.

Evaluation results Table 2 compares the results returned by TABDUAL and the ABDUAL meta-interpreter for queries to the ground programs in Figure 6.

Queries	TABDUAL	ABDUAL meta-interpreter
p_0	$[a], [b]$	$[a], [b]$
not p_0	$[not\ a, not\ b]$	$[not\ a, not\ b], [not\ a]$
not p_1	$[\ ]$	$[\ ]$
q_1	$[\ ]$	$[\ ]$
p_2	no	no
not p_2	$[\ ]$	$[\ ]$
p_3	$[\ ]$ undefined	$[\ ]$
not p_3	$[\ ]$ undefined	$[\ ]$
p_4	$[not\ a, not\ b]$	$[not\ a, not\ b]$
not p_4	$[a], [b]$	$[a], [b], [a, b]$
p_5	$[\ ]$ undefined	no
not p_5	$[\ ]$ undefined	$[\ ]$
p_6	$[\ ]$ undefined	$[\ ]$
not p_6	$[\ ]$ undefined	no
p_7	no	no
not p_7	$[\ ], [not\ a], [not\ b], [not\ a, not\ b]$	$[\ ], [not\ a], [not\ b], [not\ a, not\ b]$
q_8	$[\ ], [not\ a], [b]$	$[not\ a], [b]$
not p_8	$[\ ], [not\ a], [b]$	$[not\ a], [b]$
p_{10}	$[a]$ undefined	$[a]$
not p_{10}	$[not\ a], [a]$ undefined	$[a], [not\ a]$
p_{11}	$[a]$	$[a]$
not p_{11}	$[not\ a]$	$[not\ a]$
not q_{11}	$[a], [not\ a]$	$[\ ], [a], [not\ a]$
p_{12}	$[a]$	$[a]$
not p_{12}	$[not\ a]$	$[not\ a]$
not q_{12}	$[a], [not\ a]$	$[\ ], [a], [not\ a]$

Table 2: Comparison of results: TABDUAL vs. ABDUAL meta-interpreter

B.2 Programs with variables

Collection of programs Figure 7 lists programs with variables; many of them contain loops as well. In this collection, $a/1$, $b/1$, and $c/1$ are abducibles.

Evaluation results Table 3 presents the evaluation results returned by TABDUAL and the ABDUAL meta-interpreter for queries to programs in Figure 7.

$$p_0(X) \leftarrow q_0(X).$$
$$p_0(1) \leftarrow a(1).$$
$$q_0(X) \leftarrow p_0(X).$$
$$q_0(2) \leftarrow a(2).$$

$$p_1(X) \leftarrow not\ q_1(X), r_1(X).$$
$$r_1(X) \leftarrow not\ q_1(X), p_1(X).$$
$$q_1(X) \leftarrow s_1(X), not\ p_1(X).$$
$$s_1(1).$$

$$p_2(X) \leftarrow q_2(X).$$
$$q_2(X) \leftarrow r_2(X).$$
$$r_2(X) \leftarrow p_2(X).$$

$$p_3(1) \leftarrow q_3(1).$$
$$q_3(X) \leftarrow not\ r_3(X).$$
$$r_3(X) \leftarrow p_3(X).$$

$$p_4(X) \leftarrow q_4(X).$$
$$q_4(X) \leftarrow p_4(X).$$
$$q_4(1) \leftarrow not\ a(1), not\ a(2).$$

$$p_5(X) \leftarrow q_5(X).$$
$$q_5(X) \leftarrow t_5(X), not\ r_5(X).$$
$$r_5(X) \leftarrow t_5(X), not\ s_5(X).$$
$$s_5(X) \leftarrow p_5(X).$$
$$t_5(1).$$

$$p_6(X) \leftarrow t_6(X), not\ q_6(X).$$
$$q_6(X) \leftarrow r_6(X).$$
$$r_6(X) \leftarrow s_6(X).$$
$$s_6(X) \leftarrow t_6(X), not\ p_6(X).$$
$$t_6(1).$$

$$p_7(X) \leftarrow s_7(X), not\ q_7(X), r_7(X).$$
$$r_7(X) \leftarrow t_7(X), not\ q_7(X), p_7(X).$$
$$q_7(X) \leftarrow not\ p_7(X), not\ r_7(X).$$
$$s_7(1) \leftarrow a(1).$$
$$t_7(1) \leftarrow b(1).$$

$$p_8(X) \leftarrow s_8(X), not\ q_8(X).$$
$$q_8(X) \leftarrow not\ p_8(X).$$
$$q_8(2) \leftarrow a(2).$$
$$s_8(1) \leftarrow a(1).$$
$$s_8(2) \leftarrow a(2).$$

$$p_{10}(X) \leftarrow s_{10}(X), not\ q_{10}(X).$$
$$q_{10}(X) \leftarrow p_{10}(X), a(X).$$
$$s_{10}(1) \leftarrow a(1).$$

$$p_{11}(X) \leftarrow s_{11}(X), not\ q_{11}(X).$$
$$q_{11}(X) \leftarrow p_{11}(X), not\ a(X).$$
$$s_{11}(1) \leftarrow a(1).$$

$$p_{13}(X) \leftarrow r_{13}(X), not\ p_{13}(X).$$
$$p_{13}(1) \leftarrow a(1), b(1).$$
$$p_{13}(2) \leftarrow c(2).$$
$$r_{13}(1) \leftarrow a(1).$$
$$r_{13}(2) \leftarrow a(2).$$

Figure 7: Collection of Programs with Variables

Queries	Results by TABDUAL[a]
$p_0(X)$	$[a(1)]$ for $X = 1$; $[a(2)]$ for $X = 2$
$q_0(X)$	$[a(1)]$ for $X = 1$; $[a(2)]$ for $X = 2$
$not\ p_0(X)$	$[not\ a(1), not\ a(2)]$ for $X = \_$
$not\ q_0(X)$	$[not\ a(1), not\ a(2)]$ for $X = \_$
$q_1(X)$	$[\ ]$ for $X = 1$
$not\ q_1(X)$	no
$not\ p_1(X)$	$[\ ]$ for $X = \_$
$p_2(X)$	no
$not\ p_2(X)$	$[\ ]$ for $X = \_$
$p_3(X)$	$[\ ]$ *undefined* for $X = 1$
$not\ p_3(X)$	$[\ ]$ *undefined* for $X = \_$
$p_4(X)$	$[not\ a(1), not\ a(2)]$ for $X = 1$
$not\ p_4(X)$	$[a(1)], [a(2)]$ for $X = \_$
$p_5(X)$	$[\ ]$ *undefined* for $X = 1$
$not\ p_5(X)$	$[\ ]$ *undefined* for $X = \_$
$p_6(X)$	$[\ ]$ *undefined* for $X = 1$
$not\ p_6(X)$	$[\ ]$ *undefined* for $X = \_$
$p_7(X)$	no
$not\ p_7(X)$	$[a(1)], [not\ a(1)], [a(1), b(1)], [a(1), not\ b(1)]$ for $X = \_$
$p_8(X)$	$[a(1)]$ *undefined* for $X = 1$
$not\ p_8(X)$	$[a(1)], [a(2)], [a(1), a(2)], [not\ a(1), not\ a(2)]$ for $X = \_$
$p_{10}(X)$	$[a(1)]$ *undefined* for $X = 1$
$not\ p_{10}(X)$	$[a(1)]$ *undefined*, $[not\ a(1)]$ for $X = \_$
$p_{11}(X)$	$[a(1)]$ *undefined* for $X = 1$
$not\ p_{11}(X)$	$[not\ a(1)]$ *undefined* for $X = \_$
$q_{13}(X)$	$[a(1), not\ b(1)]$ for $X = 1$; $[a(2), not\ c(2)]$ for $X = 2$
$not\ q_{13}(X)$	$[a(1), b(1)], [a(2), c(2)], [not\ a(1), not\ a(2)]$ for $X = \_$
$not\ p_{13}(X)$	$[not\ b(1), not\ c(2)], [not\ a(1), not\ c(2)]$ for $X = \_$

[a]Underscore (_) denotes some variable, for instance in $X = \_$ (i.e. X is left uninstantiated).

Table 3: Evaluation results of Programs with Variables

Cut-free Proof Systems for Geach Logics

MELVIN FITTING
Department of Mathematics and Computer Science (Emeritus)
Lehman College
Bronx, NY

Abstract

Prefxed tableaus for modal logics have been around since the early 1970s, and are quite familiar by now. Rather recently it was found that they were dual to nested sequents, which have a complicated history but which also trace back to the 1970s. Both have provided very natural proof systems for the most common modal logics, including those in the so-called modal cube. In this paper we add some simple machinery to both prefxed tableaus and to nested sequents, producing cut-free proof systems for all logics axiomatized by Geach formulas, that is, by axiom schemes of the form $\Diamond^k \Box^l X \supset \Box^m \Diamond^m X$. This again provides proof mechanisms for the modal cube, but mechanisms of a different nature than usual. But further, it provides proof mechanisms for an infnite family of modal logics, and does so in a modular way with a clear separation between logical and structural rules. The version of nested sequents presented here has a direct relationship with the formal machinery of (Negri 2005), and can be thought of as a notational variant of a natural and interesting fragment of what can be handled using that methodology.

Meta-level Abduction

Katsumi Inoue
Inference and Learning Group
National Institute of Informatics
Japan

Abstract

Meta-level abduction (MLA) has been proposed as a method to abduce missing laws in completing proofs and explaining observations at the meta-level. Based on a simple logic of causality, (Inoue et al 2010) firstly proposed meta-level abduction to discover physically unobserved causality in terms of hidden rules to explain given empirical rules with respect to skills for music playing. Meta-level abduction has also been applied to completion of biological networks containing both positive and negative causal effects (Inoue et al. 2013). In this paper, we define a general framework for meta-level abduction together with a logical system for it, and analyze its potential power in various patterns of abductive reasoning. We will see that meta-level abduction can realize second-order existential abduction by (Schurz 2008). Moreover, meta-level abduction can be coordinated with selective, creative and other types of abductions.

Abduction and Beyond

Ari Saptawijaya*
Centro de Inteligência Artificial (CENTRIA)
Departamento de Informática, Faculdade de Ciências e Tecnologia
Universidade Nova de Lisboa, 2829-516 Caparica, Portugal
`ar.saptawijaya@campus.fct.unl.pt`

Luís Moniz Pereira
Centro de Inteligência Artificial (CENTRIA)
Departamento de Informática, Faculdade de Ciências e Tecnologia
Universidade Nova de Lisboa, 2829-516 Caparica, Portugal
`lmp@fct.unl.pt`

Abstract

In this paper we emphasize two different aspects of abduction in Logic Programming (LP): (1) the engineering of LP abduction systems, and (2) application of LP abduction, complemented with other non-monotonic features, to model morality issues. For the LP engineering part, we present an implemented tabled abduction technique in order to reuse priorly obtained (and tabled) abductive solutions, from one abductive context to another. Aiming at the interplay between LP abduction and other LP non-monotonic reasoning, this tabled abduction technique is combined with our own-developed LP updating mechanism âĂŞ the latter also employs tabling mechanisms, notably incremental tabling of *XSB Prolog*. The first contribution of this paper is therefore a survey of our tabled abduction and updating techniques, plus further development of our preliminary approach to combine these two techniques. The second contribution of the paper pertains to the application part. We formulate a LP-based counterfactual reasoning, based on *Pearl*'s structural theory, via the aforementioned unified approach of our LP abduction and updating. The formulation of counterfactuals allows us to subsequently demonstrate its applications to model moral permissibility, according to the Doctrines of Double and Triple Effect, and to provide its justification. The applications are shown through classic moral examples from the literature, and tested in our prototype, *Qualm*, an implementation reifying the presented unified approach.

---

*Affiliated with Faculty of Computer Science at Universitas Indonesia, Depok, Indonesia.

www.ingramcontent.com/pod-product-compliance
Lightning Source LLC
Chambersburg PA
CBHW081014040426
42444CB00014B/3200